Lancaster's Line to the Sea

A History of the Glasson Branch of the LNWR

Dave Richardson

Cumbrian
Railways
Association

Other publications by the Cumbrian Railways Association

An Introduction to Cumbrian Railways *(David Joy)*

'Cumbrian Communities' Series

No 1 Grange-over-Sands *(Leslie R Gilpin)* — Out of print
No 2 Ravenglass *(Peter van Zeller)* — Out of print
No 3 Dalton-in-Furness *(Rock Battye)*
No 4 Whitehaven *(Howard Quayle)* — Out of print
No 5 Millom *(Alan Atkinson)*

'Railway Histories' Series

The Coniston Railway (second impression) *(Dr Michael Andrews & Geoff Holme)*
The Track of the Ironmasters *(W McGowan Gradon — edited by Peter Robinson)*
The Ulverstone and Lancaster Railway *(Leslie R Gilpin)*
The Pilling Pig — A History of the Garstang & Knott End Railway *(Dave Richardson)*
Bigrigg — A History of the Bigrigg branch of the Whitehaven, Cleator & Egremont Railway *(The Bigrigg Research Team)*
Kendal Tommy — A History of the Arnside–Hincaster Branch *(Dave Richardson)* — Out of print

Photographic Albums

A Cumbrian Railway Album *(Leslie R Gilpin)* — Out of print
A North Lancashire Railway Album *(Leslie R Gilpin)*

Other Titles

Railwaymen of Cumbria Remembered — A Roll of Honour

Copies of books can be ordered *via* the Cumbrian Railways Association website at cumbrianrailways.org.uk

Published by the Cumbrian Railways Association,
104 Durley Avenue, Pinner, Middlesex. HA5 1JH

The Association is Registered Charity No 1025436
www.cumbrianrailways.org.uk
Please see the website for details of membership and an application form to join

Design and layout by Michael Peascod
Printed by The Amadeus Press Ltd., Cleckheaton
ISBN 978-0-9570387-9-0

Front Cover
It's the 11th March 1964 and 2MT No 46422 carries out some shunting at the District Engineer's yard before continuing down the branch. The locomotive at the rear is withdrawn Fowler tank No 42301 being taken to the ship breaking facility at Glasson Dock for scrapping. This was operated by a company called Lacmots Ltd, which rented two berths on the river quay at Glasson. Initially this operation provided some traffic for the branch but subsequently much of the scrap metal left the port by ship. On the left-hand side is a small part of the extensive Lune Mills site operated for many years by James Williamson & Son.

Derrick Codling.

Contents

A small sailing vessel discharges china clay or whiting at the western end of the Ford Quay. It is very likely that this is destined for James Williamson and Sons' St George's Works, situated only a short distance away. In the background the LNWR's Carlisle bridge carries the main line from Euston, northwards, while across the river on the left, the route of the Midland Railway's line to Morecambe is just visible. Taken around the end of the nineteenth century.

CRA. Thomas Rathbone *via* Helen Burrow.

The Cumbrian Railways Association (CRA) is a registered charity founded in 1976. Our aims are to promote and improve the knowledge and understanding of the railways of Cumbria and their social and physical impact from when they started right up to the present day. The CRA maintains written and photographic archives of much of the railways of the region, along with associated historical events. We have over 530 members, UK and world wide, and are always keen to welcome more. For more information about the Association and membership, please visit our website at www.cumbrianrailways.org.uk.

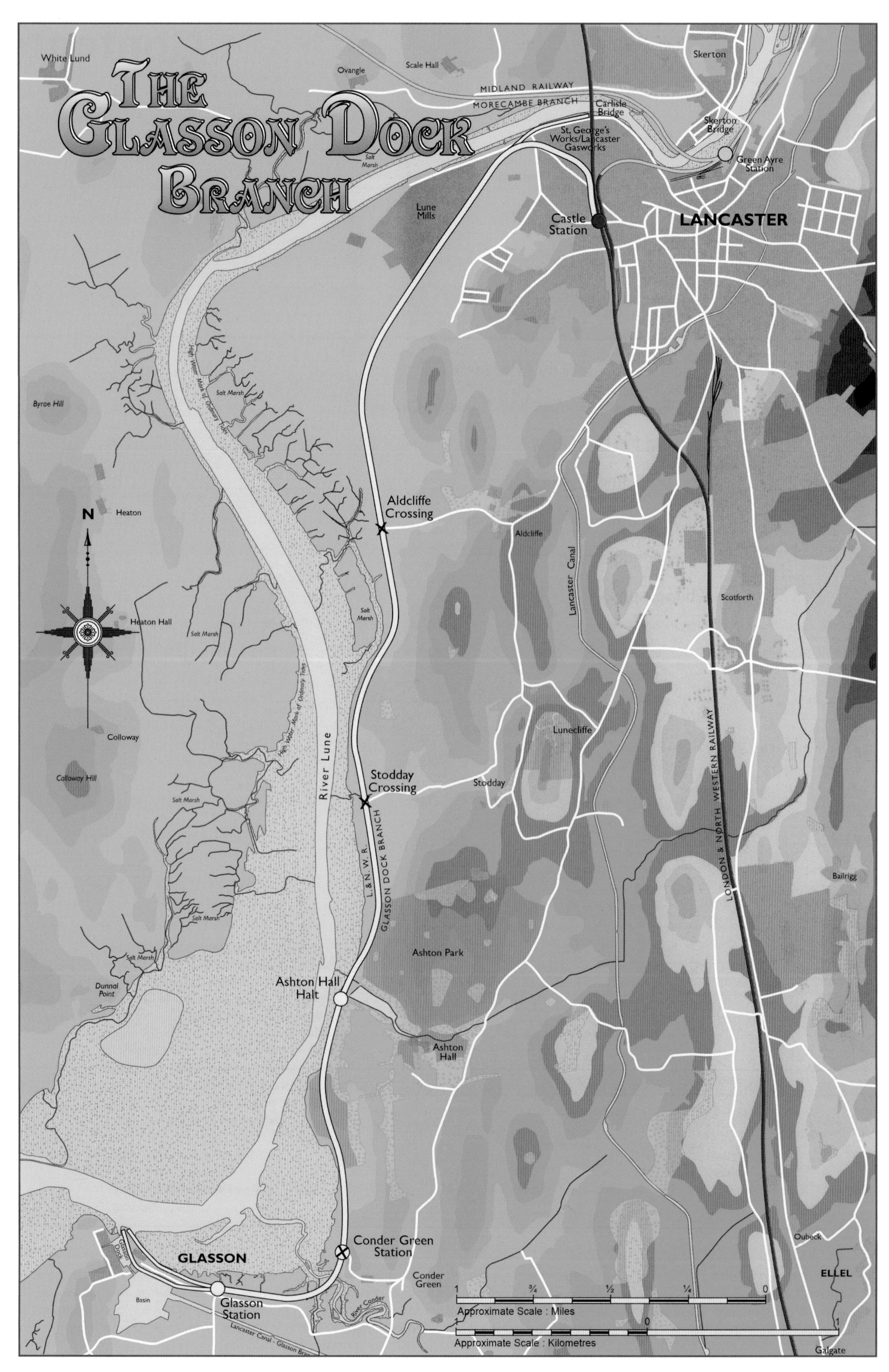
The Glasson Dock Branch
White Lund
Ovangle
Scale Hall
Skerton
MIDLAND RAILWAY
MORECAMBE BRANCH
Carlisle Bridge
Skerton Bridge
St. George's Works/Lancaster Gasworks
Green Ayre Station
Salt Marsh
Lune Mills
Castle Station
LANCASTER
Byroe Hill
High Water Mark of Ordinary Tides
N
Heaton
Aldcliffe Crossing
Aldcliffe
Lancaster Canal
Scotforth
Heaton Hall
Colloway
Colloway Hill
Lunecliffe
River Lune
Stodday Crossing
Stodday
L. & N. W. R.
GLASSON DOCK BRANCH
LONDON & NORTH WESTERN RAILWAY
Bailrigg
Ashton Park
Dunnal Point
Ashton Hall Halt
Ashton Hall
Conder Green Station
Conder Green
GLASSON
Glasson Dock
Basin
Glasson Station
River Conder
Lancaster Canal - Glasson Branch
Oubeck
ELLEL
Galgate
Approximate Scale : Miles
Approximate Scale : Kilometres

Introduction

IN LIVERPOOL, A PROUD CITY steeped in maritime history, there is an old saying, half remembered in families with some former connection to the sea: "If you don't behave, I'll send you to Glasson Dock". It was used, apparently, as a threat to naughty children and is surely a slur on what is today an attractive and interesting village and a popular destination for visitors coming in by car, motorbike and bicycle. However, just for a moment, put yourself in the place of a young sailor from the teeming commercial metropolis of Victorian Liverpool, arriving at Glasson Dock, say around 1875. In favourable conditions, it is perhaps only a day's sail away. He would arrive at a small wet dock on a lonely, isolated headland at the mouth of the River Lune. Other than the three pubs and a few rows of cottages huddled around the dock and the adjoining canal basin, the only other signs of nearby habitation are a handful of scattered farms and cottages. The large town of Lancaster is only five miles away but is completely hidden around the bends of the River Lune. The only links with the outside world are a narrow road disappearing across the salt marsh and the Lancaster Canal running arrow straight across the flat open fields. Yet behind this unprepossessing first impression for our young Victorian Liverpudlian, lies the story of the port and of the railway that served it, which are both of considerable interest.

The branch line to Glasson Dock was constructed by the London and North Western Railway (LNWR) from their Lancaster Castle Station, to this small port at the mouth of the River Lune. Lancaster itself was over 200 miles from the railway's London terminus and headquarters at Euston Station. However, Castle Station, in the past, had been the headquarters of the old Lancaster and Carlisle Railway. This was undoubtedly the most dramatic and demanding section of the LNWR's main line between London and Carlisle. Arguably, it was also the most interesting. A traveller intending to visit Glasson Dock from the south in the late 1920s would have to alight at Castle Station and then walk just a few steps across to the bay platform to board the train to his or her destination. At this point in the branch's history, the train would be composed of three, six-wheeled carriages headed by a 2-4-2 tank engine. As befits a passenger service on a very minor branch line, this is an older locomotive, and the six-wheeled carriages are definitely from another era. However, the entire ensemble is painted in the attractive, lined out crimson livery adopted by the LNWR's successor, the London Midland and Scottish Railway. Immediately after departing from the bay platform the train leaves the main line and curves down to the floodplain of the River Lune. Once the industry of Lancaster is left behind, the line crosses the flat windswept countryside and then runs for several miles alongside the river estuary to reach the small passenger station on the edge of the village. Here, the line divides, to make two end-on connections with the Port Commissioners' own sidings that served the dock facilities.

Apart from the story of the branch line, running through this narrative like a thread, is the story of the port itself. This has a rich and interesting history deserving a book all of its own and it would undoubtedly make for a fascinating story. However, given that I was supposed to be writing about the railway, I have hopefully been judicious in including only material supporting the railway narrative, as opposed to running parallel to it. Nevertheless, there is rather more about shipping matters than would normally feature in a book about a railway branch line. I pondered on this whilst putting the book together, but concluded that it was simply not possible to fully understand the story of the railway line without having a good grasp of what was also happening at Glasson Dock and, indeed, at the quays further up the river at Lancaster to the industries that clustered either around the dock at Glasson or at the Lancaster end of the branch. Over time, the railway's principal customer was the firm of James Williamson & Son, who were linoleum manufacturers *par excellence*, with large markets both at home and abroad, and whose colossal Lune Mills sat alongside the line a little to the south-west of Lancaster. James Williamson, later Lord Ashton, used a changing and almost bewildering mixture of rail, sea and, later on, road transport to bring in raw materials. On the other hand, the large traffic in finished products invariably left the works by rail. Making sense of all of this, is one of the keys to understanding the traffic patterns prevalent on the branch at different times.

One unconventional aspect of this book (for a railway book, that is) is the extensive use made of the records of the Lancaster Port Commissioners. These are held at Lancashire Archives in Preston and I spent many enjoyable hours delving into this substantial deposit. Paradoxically perhaps, the minutes of the Commissioners' monthly meetings chart the history of the branch line in far greater detail and intimacy than any railway company records ever could. However, their use has undoubtedly given the narrative a different, customer-based perspective. It is the view from the other side of the railway fence so to speak. Using these records has also provided a distinctly local viewpoint, something which is more difficult to obtain from the records of a large railway company. However, perhaps the most desirable consequence of utilising the Commissioners' records was that they provided a great deal of information about the interface between the operation of the railway and the functioning of the port of Lancaster as a whole. As far as I could recall, this was not an aspect of railway history I'd previously seen examined in any great detail, at least not for the docks and harbours of North West England. I also felt that rather than getting caught up in the complex operations at a large port, it would be interesting and instructive to see how this played out on a much smaller scale at Glasson Dock.

Much of the success of any port depends upon efficient transport links and the ability of the operators to load or discharge cargoes as quickly as possible. At Glasson Dock, both of these aspects were, until 1938, largely in the hands of the LNWR and its successor, the LMS. Local goods traffic on the branch was normally relatively light. However, from time to time, it was necessary to procure a large number of suitable wagons, sometimes at short notice, in order to load a ship's cargo of, say, iron ore or wood pulp. The wagons would have to have been loaded quickly and then dispatched to their destination with the least possible delay. How well, or how badly the railway performed these operations is one of the themes of the story of the branch.

Whilst seeking material for this book, one of the most interesting discoveries was the existence of two separate letters from Mr EM Gilmour who, between 1927 and 1930, was the stationmaster at Glasson Dock. In them, he recounts his experiences and provides a unique insight into many aspects of the railway's operation during this period. But for the chance survival of these items of correspondence, much of this detail would have been lost. Another good "find", although hidden within plain sight, was the photograph collection at Lancaster's Maritime Museum. In connection with this, I had two very productive and enjoyable visits to their premises, situated, very appropriately in the Custom House on St George's Quay, going through their collection of shipping photographs of Glasson Dock and the quays at Lancaster. These images repaid careful study, as, half hidden in the background of some of them, were items of significance and interest to the railway enthusiast and historian. Some of these images appear in this book. The information contained in them helped to fill in some of the gaps in the narrative or provided confirmation of what had hitherto only been surmised. The use of archived editions of Lloyds Register (available on the internet) in conjunction with the Port Commissioner's shipping registers made it possible to date some of the photographs and revealed what was being discharged into the railway wagons standing on the quay side.

I very much enjoyed researching and writing this history of a railway line that has fascinated me since childhood. Some of the images I came across, taken in the early 1960s, brought to mind Summer Sunday excursions to Glasson Dock with my parents in our Norton Atlas motorcycle combination with its two-seater Watsonian sidecar. This project has also prompted me to delve deeper into the history of shipping on the Lancashire coast and also perhaps, to get a little more in touch with my inner, salty sea dog.

Dave Richardson

Prescot, March 2022

Chapter One

Before the Railway

Glasson Dock before the coming of the railway. This is a slightly unusual view taken from the foreshore alongside the entrance to the wet dock. The building in the centre background is the grain warehouse, which dated from the opening of the canal to Glasson, and was demolished in 1939.
Courtesy Lancaster City Museums.

THE VILLAGE OF GLASSON DOCK owes its existence to the construction first of all of a pier and then an enclosed wet dock at a lonely spot close to the mouth of the Lune estuary on the Lancashire coast. However, to understand how this came about, it is necessary to look towards Lancaster, some five miles upstream. Towards the end of the eighteenth century, the county town of Lancaster was a busy river port with a vigorous Irish Sea trade, as well as a significant international trade with the West Indies and the Baltic. At this point, it ranked as one of the principal ports in North West England, although still a long way behind the much larger port of Liverpool to the south.

As trade increased and, with it, the size of the ships used on the longer haul voyages, the limitations of the winding, shallow River Lune channel became apparent. This problem exercised the minds of the Lancaster Port Commissioners, a statutory body authorised by an Act of Parliament in 1749 and tasked with the administration and improvement of the Lune navigation for the benefit of the traders and shipowners using the port. In the years from 1750, the Commissioners constructed St George's Quay on the south bank of the river just below the castle and, by 1759, there was a quay wall of 200 yards in length with wharves and warehouses. The centrepiece was undoubtedly the Palladian-style Custom House, which was completed in 1764 and today is the home of Lancaster's Maritime Museum. However, as trade increased and ships inevitably became larger, the largest could only reach the quay at Lancaster with some difficulty. As a consequence of this, it became the practice to discharge these larger ships at Sunderland Point on the northern side of the Lune estuary and then to cart the cargoes overland to Lancaster. This created difficulties of its own, not the least of which was the fact that the road between the two places was regularly flooded at high tides. A further manifestation of the necessity to cater for larger vessels was the construction, in 1768, of a second quay, just a little further downstream from the town, where it was hoped to take advantage of the increased depth of water in the river channel. Because the construction post-dated that of St George's Quay, this second facility was always known as the New Quay

However, the Commissioners decided that the long-term answer was to construct a river quay and an adjacent wet dock at Glasson, across the estuary from Sunderland Point and about six miles downstream from Lancaster itself. The river quay was brought into use in 1781–2 and the wet dock itself was finally opened on 20th May 1787. Although the new facilities were clearly beneficial to the trade of Lancaster, in the longer term, they appear to have done little to arrest the general decline in Lancaster's position as a significant port, which took place after 1800. The reasons for this decline are complex, but one important factor was the general dislocation of trade and finance, which took place during long-running Napoleonic Wars. This, in its turn, was exacerbated by the trade depression that set in following the cessation of hostilities in 1815. These difficult conditions tended to push ship owners, merchants and even local industries dependant on imported materials, towards the much larger port of Liverpool with its growing industrial hinterland.

The next major development at Glasson Dock was the construction of a branch of the Lancaster canal from Galgate, a small village a little to the south of Lancaster. The original powers to construct a canal from Westhoughton (between Bolton and Wigan) to Kendal, together with a branch to Glasson Dock, were authorised in an Act of Parliament in 1792. Additional powers for a branch canal to Glasson Dock and the sea were obtained *via* a second Act in the following year. The vicissitudes of the Lancaster Canal project are outside the scope of our story, but in the event, an isolated section of the canal between Preston and Tewitfield, just north of Carnforth, was opened in 1797. However, it was not until 18th June 1819 that the final 14-mile-long section northwards to Kendal was opened. Unfortunately, the route between Preston and Kendal was destined to remain isolated from the rest of the national canal network. Having completed this though, thoughts now turned inevitably to the construction of the branch to Glasson Dock on the coast, originally authorised back in 1792. This was finally completed and opened for traffic in 1826. The branch terminated at Glasson in a large basin covering an area of some 17 acres. Access to the wet dock was *via* a lock, 100 feet long and 26 feet wide. A large grain warehouse was erected at the basin and this was designed so as to allow barges to enter and load or unload under cover. Because the canal between Preston and Kendal was completely isolated from the rest of the network, it became the practice to employ vessels with removable masts, which could operate on the canal or at sea with equal facility. These would either sail from Liverpool or enter the River Douglas *via* the Rufford branch of the Leeds and Liverpool canal. They would sail up the coast to Glasson Dock and then enter the Lancaster Canal *via* the wet dock. In this way, for a little while at least, Glasson became something of a minor transport hub. Significant amounts of coal from the Wigan area were brought to Lancaster and Kendal *via* the Leeds and Liverpool and Lancaster canals, with the short sea voyage to Glasson providing a link between the two. Additionally, commodities from abroad were brought directly to Glasson Dock and transferred from large ocean-going vessels into canal barges to be taken on to their destination. For example, it is known that at least one timber merchant in Preston received consignments from Canada *via* Glasson Dock and the Lancaster Canal. The dock dues paid by vessels visiting the wet dock or passing through onto the canal, placed funds in the hands of the Lancaster Port Commission, which could be utilised for further improvements to the dock itself and to the river navigation.

Although comparative figures are not readily available, this period from the opening of the canal perhaps through to the 1840s must have represented the pinnacle of Glasson's relative prosperity and importance. However, by the middle of the nineteenth century, further developments in the transport infrastructure of the wider area, are likely to have reduced the amount of traffic coming into the port. Between 1839 and 1843, the channel of the River Ribble, 20 miles to the south, was significantly improved and new quays were constructed at Preston. By early 1846, these were connected to the railway network by means of a short steeply-graded branch from the station. Nearer to Glasson Dock, a new town and a harbour were established at the mouth or the River Wyre. This was named Fleetwood, after its founder, Sir Peter Hesketh Fleetwood. The Preston and Wyre Railway, connecting the new town with the rest of the railway network, was opened on 15th July 1840. By the Summer of 1846, there was a total of 2,400 feet of quay frontage on the river at Fleetwood. Even closer at hand at Morecambe, a harbour was developed during the early 1850s by the North Western Railway. This company, not to be confused with the London and North Western Railway, had, by 1850, completed a route from the West Riding of Yorkshire into Lancaster and Morecambe. The railway provided cranes for the loading and unloading of ships and storage facilities for discharged cargoes.

Each of these new port facilities had the undisputed benefit of being served by a railway. However, it was the expansion of the railways themselves that had perhaps the biggest impact on the fortunes of Glasson Dock. The first line to penetrate the area was the Lancaster and Preston Railway, which was opened on the 25th June 1840. This was followed, on 21st September 1846, by the opening of the Lancaster–Oxenholme section of the Lancaster and Carlisle Railway. From 1st August 1859, both the Lancaster and Preston and Lancaster and Carlisle Railways were leased to the London and North Western Railway for 900 years. However, the company had assumed effective control of the whole route between Euston Station in London and Carlisle, some time before this. As we shall see, the LNWR was to have a significant influence on the future of Glasson Dock. Finally, there were the lines of the North Western Railway which have already been touched upon in connection with the construction of the harbour at Morecambe. The branch between here and the station at Lancaster Green Ayre was opened on the 12th June 1848. The route eastwards into Yorkshire was completed in stages with the final section being opened for traffic on 1st June 1850. The Derby-based Midland Railway took over the operation of the North Western Railway on 1st June 1852, finally leasing it for 999 years with effect from 1st January 1859.

Lancaster Castle Station looking north before it was remodelled in the period 1898–1902. The single bay platform for the Glasson Dock passenger trains was situated through the overbridge on the left-hand side.

Lens of Sutton Association.

An early view of St George's Quay showing the original Lancaster and Carlisle Railway bridge. The arches were constructed in timber and, by the early 1860s, were proving inadequate for the increasing weight of traffic. By July 1865, plans had been drawn up to replace them with wrought iron main and cross girders. At the end of that year, tenders were invited for the work and the contract was awarded to the Fairburn Engineering Co for £19,926.

Courtesy Lancaster City Museums.

So, by 1850, Lancaster had rail routes leaving the town on all four points of the compass. In that same year, the Lancaster and Carlisle Railway and the Lancaster Canal Co. had come to an agreement whereby the traffic from the south was split between them, the canal carrying the coal and heavy bulk materials, and the railway carrying general merchandise and passengers. However, this comfortable arrangement came to an end in 1858, when the railway began challenging the canal for the lucrative Wigan coal traffic. Faced with a significant loss of revenue, the Lancaster Canal Co was inexorably pushed into the arms of its railway rival which, by August 1859, had become, effectively a part of the LNWR. In November 1862, after over two years of negotiations, the LNWR promoted a bill for the lease, sale or transfer of the canal undertaking and its property, to themselves. The first bill was unsuccessful, but undeterred, the LNWR came back with a second bill for the following parliamentary session and on this occasion they were successful. The Lancaster Canal Transfer Act , receiving royal assent on 29th July 1864, sanctioned the leasing of the canal from Preston northward, to the LNWR, in perpetuity, for an annual rent of £12,665 17s. 6d. This arrangement lasted until 1st July 1885, when the LNWR purchased the canal outright.

Given the growth of the railway system and the increased competition from other port facilities in the area, all of them with rail connections, it must have become apparent to the Lancaster Port Commissioners that their own facilities on the River Lune at Lancaster and at Glasson Dock were operating at something of a disadvantage. Admittedly, the quays at Lancaster were not too far from the Midland Railway's station at Green Ayre, just a little further up the river, and they were even closer to the LNWR's Castle Station. However, there was no rail access from either, which was essential if discharged cargoes were to be transferred directly into railway wagons. At one point, the North Western Railway had considered a scheme to put in a line from their station to run along the south bank of the River Lune and onto the quays at Lancaster. However, this was not proceeded with. Out at the mouth of the river estuary, Glasson Dock itself was several miles away from the railway network. However, almost paradoxically, the port now had a significant link with the LNWR by virtue of the latter's position as lessee of the Canal Co. Access to the Lancaster Canal, of course, did continue to channel some trade through the port, but by 1870 any port connected to its hinterland solely by a canal had become something of an anachronism. Clearly, if Glasson Dock was to develop and thrive, some sort of connection with the national railway network was essential.

Chapter Two

The Origins of the Scheme

ALTHOUGH IT SEEMS to have been clear to the Port Commissioners, and indeed to the LNWR, that more could be done to develop the facilities at Glasson, it was several years before their plans began to align and coalesce into positive action. Such plans were to the ultimate benefit of both parties, of course, although their perspectives were different. For the LNWR, one of the largest joint stock companies in Britain, Glasson Dock was merely a far-flung outpost of its Euston-based business empire. The company already leased the Lancaster Canal, of course, a branch of which served the port and, by 1870, there was still a significant amount of traffic passing between the two. However, with the correct development and suitable incentives for traders to use the dock, a railway line connecting the port with the main network might prove a useful additional source of income. For their part, the prominent businessmen who constituted the bulk of the Port Commissioners must have seen that the provision of a railway line to the port was essential if trade was to prosper in the future. There was also another aspect to this. For the port of Lancaster as a whole, a significant amount of business was still carried on at St George's Quay and the New Quay both on the south side of the river at Lancaster. When the Lancaster and Carlisle Railway was constructed, the Carlisle bridge, carrying the line over the river, effectively cut St George's Quay into two. In response to this, the Port Commissioners constructed a timber-faced extension onto the front of the existing quay, immediately to the west of the railway bridge. This extended the whole area a little further out into the river. After the coming of the railway, coasting vessels continued to negotiate the winding river channel upstream of Glasson Dock, to discharge their cargoes either at the New Quay, or at the upgraded St George's Quay just below the Carlisle railway bridge. If this part of the wider port's facilities could also be served by a railway connection, then so much the better.

In January 1871, the Port Commissioners learned that the LNWR was considering the construction of a branch railway to Glasson Dock. The Commissioners' minutes indicate that this information came to them informally, *via* a Mr Fitsimons. This is likely to have been John Fitsimons, the Goods Manager of the Lancaster and Carlisle section of the LNWR. He was based at Lancaster's Castle Station and, therefore, would have been known to most of the businessmen who constituted the body of the Commissioners. At a subsequent meeting convened to discuss the matter, the Commissioners decided that they would afford the LNWR every possible means to enable them to carry this out, and that a committee would be formed to confer with the railway company on this matter. In spite of the Commissioners' best intentions, and their undoubted enthusiasm for the project, nothing further seems to have come of this, at least not for the moment. In November of 1871, the Commissioners decided to investigate the possibility of having a siding or tramway running along the river, serving both St George's and the New Quays and connecting with the Midland Railway's station further upstream at Green Ayre. Whilst not serving Glasson Dock itself, such a tramway would go at least some of the way towards linking the port with the railway network. However, this was by no means a new initiative. In 1851 and again in 1854 the Commissioners had applied to Lancaster Corporation for permission to construct such a tramway. This had been readily granted subject to certain provisions relating to the crossing of some of the roads on the route. However, the work was not proceeded with. There is no evidence that the Midland Railway was involved in this most recent proposal and as in 1851 and 1854, the matter does not appear to have been pursued any further.

Here matters rested until January 1874, when two of the Commissioners were tasked with approaching the LNWR to see if they would be interested in enlarging the lock between the wet dock and the canal basin. This had nothing to do with any railway scheme for Glasson Dock of course. Rather it was part of an initiative to improve the facilities at the port. Again, this was not a new idea. Back in 1861, there had been discussions between the Commissioners and the Canal Co on this very issue. Unfortunately, these came to nothing, as the two parties could not come to an agreement as to the extent to which the lock should be enlarged. The details of the 1874 negotiations with the railway company do not appear to have survived. However, the two Commissioners concerned reported that any application for this at the present time would be useless, suggesting that they had received a polite, but very firm rebuff from the railway company.

Nevertheless, the Commissioners' approach to the LNWR concerning the enlarging of the lock, may have prompted the latter to consider what could be done to improve things more generally at Glasson Dock. In 1875, the railway formed a special committee to further consider the enlargement of the lock but also to look at the feasibility of converting the branch canal from Galgate into a railway route to the port. The Commissioners do appear to have been aware of this development because, in November 1875, they held a special meeting to consider whether any inducements could be provided to the LNWR to persuade the company to construct a railway to Glasson Dock. Once again, it was resolved that every facility and inducement be provided and that a committee be appointed to confer with the railway company on the matter. In order to provide a little further encouragement to the Euston company, a copy of the Commissioners' resolution was forwarded to Richard Moon, the Chairman of the LNWR. However, all of this merely took the situation back to how it stood in 1871 and it appears that in the ensuing five years, little or no progress had been made towards the provision of a railway line to Glasson Dock. Nevertheless, it appears that the Commissioners had learned a little about the art of public relations. In September 1876, they arranged for 20 copies of a photograph of Glasson Dock to be produced and distributed amongst interested parties. Particular care was taken that a copy was sent to Mr Moon of the LNWR. Part of a charm offensive no doubt, and also a reminder for him of the existence of the small port at the mouth of the River Lune.

The mood in the Lancaster area was summed up by a correspondent in the *Lancaster Gazette* for the 16th September 1876. He bemoaned the absence of a railway to Glasson Dock, saying that in the past three months, 6,960 loads of timber and 5,980 tons of Indian corn had been brought into the port, all of which had had to be taken inland by the slow and inconvenient process afforded by the canal. He commented that many thousands of tons of corn were imported into Barrow, which would come to Glasson if it was possible to have it conveyed to its destination by rail. He ends his letter by saying that he hoped the LNWR would see the desirability of constructing a branch from their existing line at Galgate.

Whilst there were now faint signs that the LNWR might indeed be persuaded to build a railway to Glasson Dock, at this stage the actual route of the line was still uncertain. Furthermore, it was by no means clear as to whether any proposed line would also serve the quays at Lancaster. With this in mind, in October 1876, the Commissioners revived their scheme for a siding or tramway running from St George's Quay at Lancaster to the Midland Railway's station at Green Ayre. As usual, a committee was appointed to report on this matter. However, at this point, the Commissioners must also have been aware of a forthcoming privately-promoted bill for a railway from Galgate to Glasson Dock. The proposed line went nowhere near the quays at Lancaster and this may have been the reason for their reconsideration of the old scheme.

The statutory announcement of the parliamentary bill for this privately-promoted rail link for Glasson Dock duly appeared in the *London Gazette* and the local newspapers in mid-November 1876. The line was to commence at the northern end of the platform of Galgate station and run through

A view from around 1900 showing a small sailing vessel entering the wet dock at Glasson. On the right is one of the Port Commissioners' two steam cranes. These were run up and down this siding nearest the edge of the quay, while the wagons for loading or unloading were placed on the inner tracks. Commercial postcard, Author's collection.

the parishes of Cockerham, Ellel, Ashton with Stodday and Thurnham to Glasson Dock. Here it would terminate opposite the Victoria Hotel at a point 50 yards to the south of the eastern wall of the hotel. Powers were also to be requested to sell or lease the undertaking to either the LNWR or the Midland Railway or to both jointly. The way would also be open for either company to construct and maintain the undertaking themselves, or to subscribe to the cost, or to merely acquire the line upon completion. It seems that the aim was to create a sort of off-the-shelf railway scheme, which would be attractive to either the Midland or the LNWR. Clearly Euston was well aware of the agitation within the local area for a line to serve Glasson Dock. However, this privately-promoted scheme had the potential to see Midland trains running into Glasson Dock, which would have been unwelcome to the LNWR. They already leased the canal serving the port facilities at the village and would not have wanted to see any expansion of their business rival's interests into what they regarded as their own territory. The bill was warmly supported by the local newspapers, the *Lancaster Gazette* for 15th November 1876 averring that the proposed railway would be of immense advantage to the port and conducive to the prosperity of Lancaster.

On the 20th November 1876, the Commissioners met to receive a deputation from the promoters of the railway consisting of Mr Holden, the solicitor for the line, Mr Swettenham the engineer and a Mr M Simpson. This latter gentleman may well be Matthew Simpson, who at this time was running a successful ship-repair and ship-building business from the graving dock at Glasson. He would clearly have had a significant vested interest in any project that would connect Glasson Dock to the railway network. It is very likely, therefore, that he was at least one of the promoters of the Galgate scheme. Having heard a statement from Mr Swettenham and inspected the plan of the route of the line, the Commissioners professed themselves satisfied with what was proposed and confirmed that they would fully support the line. Clearly, this was the line that the Commissioners had been hoping for and, whilst not promoted by one of the two railway companies serving Lancaster, it nevertheless showed great promise. Having slept on the matter, the Commissioners met again on the following day. Anxious that the line should be built, they decided that they were prepared to put their hands into their pockets and provide some financial inducement. They announced that with a view to facilitating the construction of the railway, and subject to the authority of Parliament, the Commissioners were prepared to remit, to the railway company, from the opening of the line, one half of the dock and harbour dues relating to ships discharging at Glasson Dock where the cargoes subsequently passed along the railway. However, this generosity had its limits and the annual payment to the railway was not to exceed £400.

The potential threat of incursions into their territory, together with the financial concession made by the Commissioners, seem to have been sufficient to push the LNWR towards a more serious consideration of the Glasson Dock project. In early December, Messrs Holden and Swettenham were invited to Euston for an interview with Richard Moon, the chairman of the LNWR. The occasion is described in detail in a letter from Holden to the Port Commissioners, dated 11th December 1876. There was a long conversation during which the two gentlemen emphasised the importance of the undertaking to the area around Lancaster, the favourable terms made with Gerald Fitzgerald, the principal landowner along the route, and the liberal offer of the Port Commissioners. In reply, Moon was pleased to state that the scheme now appeared more desirable than he had hitherto regarded it. He said he would consult his colleagues on the LNWR board with a view to the company constructing the line themselves next year. He gave no definite undertaking that they would do this, but he did confirm that his company would not be adopting the route of the privately-promoted line, even though the survey had been completed, the plans had been prepared and the requisite parliamentary notices given. In the circumstances, the promoters of the Galgate scheme decided to withdraw their bill in order to give the LNWR the opportunity of promoting their own line the following year.

So, although the proposed Galgate route had been rejected by the LNW, it was looking increasingly likely that the company would now build a branch line to Glasson Dock themselves. In the meantime, with the precise line of the railway still uncertain, the Commissioners' Tramway Committee had been preparing their report and this was submitted in February 1877. After some discussion, it was decided to arrange for a survey of the proposed route and to have plans drawn up. Lancashire Archives in Preston holds a set of plans and sections, dated 1877, of the proposed tramway, running from the Midland station at Green Ayre and then along the edge of the river and under the Carlisle railway bridge, to a point a short distance beyond the gasworks. There were to be branches off the main route to serve the gasworks, Williamson's St George's works and St George's Quay. In addition to these, there were three other sidings serving businesses situated between the Carlisle railway bridge and Green Ayre. The survey appears to have been made entirely at the behest of the Commissioners and there is no evidence that the Midland Railway was involved, although they must surely have been consulted at some stage. In any case, the whole scheme was about to be overtaken by events and rendered obsolete.

Assuming that the LNWR was now going to build a line to Glasson Dock, the company would have to wait until November 1877 before it could submit a bill for the ensuing 1878 parliamentary session. It seems that the interim period was used to gather further information and to firm up the plans for the branch. LNWR Special Committee minute 42487, dated 19th October 1877, provides a useful and informative update:

Read letter from Mr Simpson 15 October again urging construction of a Branch Railway to Glasson Dock, & the Chairman having reported the wish of the Lancaster Gas Company to have a connection and the desire of the Lune Commissioners to have a Line of Rails along the Quays at Lancaster towards which they would be prepared to contribute some surplus funds they have in hand, & probably give some lands by the river side, which will be available for continuing the Line to Glasson without interfering with the Canal. It was referred to Messrs Hull, Worthington, Fitzsimons and Slinger to meet the Lancaster Local Authorities, Mr Simpson and others interested in the Glasson Dock, & report to the Chairman what arrangements can be made with a view of the Line being deposited for the ensuing Session. Mr Cawkwell pointed out that by crossing the road near the Quay by a bridge on the high level the gradient down to the lower level can be improved.

Mr Hull was from the LNWR estates office in Preston; Mr Worthington was the line's civil engineer; Mr Fitsimons, who we have already come across, was the Goods Manager of the Lancaster and Carlisle section of the LNWR; and Mr Slinger was from the railway's canal office in Lancaster. Finally, Mr Cawkwell was William Cawkwell who had been General Manager of the LNWR until retiring in 1874, when he joined the Board of the company.

Clearly by this time, the exact line of the proposed railway was close to finalisation. However, it was now looking very likely that the LNWR would also construct a line to serve the quay at Lancaster. It is interesting that at this early date, the Lancaster Gasworks, at this time a privately owned concern, had been in touch with the railway company requesting a siding connection. Speaking some years later, in 1893, Sir Thomas Storey, the Mayor of Lancaster and a port commissioner, stated that he and others had taken some little trouble in persuading Richard Moon and the directors of the LNWR to construct a line down to the quay at Lancaster. This may well have been the case and much of the personal contact and discussion that took place between interested parties around this time remains unrecorded in formal minutes or newspaper reports and, therefore, is hidden. The Commissioners themselves, of course, were very keen that the railway should be constructed. Consequently, they made a point of being heavily involved in the project, wishing to keep a close eye on its progress and also to be on hand to smooth out any difficulties that might arise. To this end, by the 29th October 1877, they had formed what was in effect a Glasson Dock Railway Committee to liaise with the representatives of the LNWR, the Corporation and the landowners, concerning the proposed route and its impact on the port facilities at Lancaster and Glasson Dock. The relationship between the railway company and the Port Commissioners was to be a close and, on the whole, a constructive one.

It seems that some sort of meeting with the LNWR took place very shortly after this. Lying loose in the relevant Commissioners' minute book are three faded sheets of LNWR notepaper headed *Lancaster 1st November 1877* and entitled *Heads of Agreement.* Some sections of the text are almost illegible and there are crossings out and insertions. They appear to be a set of notes taken down during a meeting between the Commissioners and the representatives of the LNWR. What is interesting is that the content mirrors very closely the terms of the formal agreement between the railway company and the Commissioners, which was eventually signed some months later. This agreement was to provide the framework for co-operation between the two parties for both the construction of the branch and for the subsequent operation of the port facilities. The detail of this will be considered in due course.

Meanwhile, later on in November 1877, the statutory notice of the LNWR's intention to submit a bill to Parliament for the construction of the railway, appeared in the *London Gazette* and the local newspapers. The LNWR was to apply for powers to construct the line in their omnibus bill for the 1878 parliamentary session. The omnibus bill was a parliamentary device used by the

A relatively rare view of a train being shunted along the line leading to the river quay at Glasson Dock. In 1963, two of the river berths were taken over by a firm of ship breakers and scrap metal merchants and the sidings were improved to serve the facility. On this occasion, withdrawn Fowler tank No 42301 (just out of sight at the front of the train) is being propelled into the sidings for disposal.

Ron Herbert.

A busy scene at Glasson Dock probably taken in the early years of the twentieth century. The structure in the foreground is the swing bridge which crosses the lock connecting the wet dock with the canal basin. The Commissioners' two steam cranes are visible on the right-hand side.
Courtesy Lancaster City Museums.

larger railway companies who, in the course of a year, might have several improvement or expansion schemes in hand. Thus, the details for the new line to Glasson Dock sat alongside proposals for work in places as diverse as Wolverton, St Helens and Salford. The proposed line was to run from Lancaster Castle Station, then after cutting across country, was to follow the southern bank of the River Lune to Glasson Dock. Additionally, there was to be a branch running onto St George's Quay at Lancaster. For the purposes of the bill, the route was divided into three sections. Section one constituted a line from Lancaster Castle station to a point on Lancaster marsh, just under a mile from the junction. Section two was for the continuation of the first section along the river to Glasson Dock. This was by far the longest of the three sections, it being a little over four miles to the termination point at the south east corner of the dock. The third and final section, and the shortest, was for the branch from section one onto St George's Quay on the River Lune.

With the whole project now out in the open so to speak, the *Lancaster Guardian* for 10th November 1877 had the following to say:

> *It seems to be now a settled thing that the plans of the London and North Western Railway for the construction of a line of railway between Lancaster and the port of Glasson will be carried out. No obstacles seem to be in the way. The landowners, whose estates the proposed line will traverse are three: Mr EB Dawson, Mr JPC Starkie MP, and the Lancaster Corporation and they are all willing to forward as far as they severally can the execution of the scheme. As at present arranged the line will commence somewhere near the Marsh Lane railway bridge, run down parallel with Marsh Lane on the North side with a branch siding to the gas works across Lune Road, through the marsh field and skirting Messrs Williamson's manufactory, across Freeman's Wood and so on through the estates of Mr Dawson and Mr Starkie, to Glasson.*

Lancaster Corporation was entirely supportive of the bill, seeing that the new line could only be of benefit to the trade of both Lancaster and Glasson Dock. However, it was concerned that the proposed line cut across certain rights of way when crossing Lancaster marsh which they were keen to preserve. Similarly, the short branch line to St George's Quay crossed over the town's outlet sewer, and the Corporation was anxious that this should be protected. This was all settled to the Corporation's satisfaction and clauses were inserted in the final Act of Parliament to protect the sewer. In the event, the land on the marsh required for the new line was transferred to the railway company by the Corporation for a perpetual rent of £8 per acre *per annum.* This was considered to be a good bargain for the LNWR and the generous terms were criticised in some quarters. However, the low rent and special terms were agreed to, as an inducement to the railway to build the branch.

On the 21st June 1878, the formal agreement between the LNWR and the Port Commissioners was signed. It was a wide-ranging document and it is worth examining the clauses in some detail. In the first instance, the LNWR was to apply to Parliament to construct the railway and the Commissioners were to use their local influence to assist with this in every way possible. This actually carried more weight than is perhaps apparent. Some of the Port Commissioners were also members of Lancaster Corporation. Indeed, the Lord Mayor was always an *ex-officio* member of the Commissioners. Once the Act received royal assent, the Commissioners, within three years, would extend the existing river quay at Glasson Dock, thereby increasing the facilities at the port. This work would be carried out by the Commissioners at their own expense. However, the land for this was to be acquired and paid for by the LNWR under the compulsory purchase powers requested in the bill. The Commissioners would subsequently refund to the railway company the cost of the land and any expenses incurred in the compulsory purchase. Upon completion of the line to Glasson Dock, the Commissioners, at their own expense, would lay sidings alongside both the wet dock and the extended river quay and provide the necessary points, crossings, turntables, cranes, hoists and other appliances required for the conduct of the railway company's traffic to and from the port. The LNWR would be free to use these sidings and appliances and, in return, would pay to the Commissioners a sum equivalent to 4.5% *per annum* of the capital cost. The railway would also refund to the Commissioners any reasonable costs incurred in maintaining the sidings and appliances. In fact, what happened in practice is that the railway company simply carried out the maintenance of these items itself. All the LNWR's traffic was to pass over the quay lines free of any toll. Also, the Commissioners were to provide storage space on the quay where this was necessary for the accommodation of the railway's traffic and, in their turn, the LNWR would pay rent for the space so provided.

At Glasson Dock, therefore, there was to be a clear distinction between the lines owned by the LNWR and running on their own land and the sidings owned by the Commissioners and running on the land belonging to the port. On St George's Quay at Lancaster there was no such distinction and the LNWR's sidings ran onto the Commissioners' property. As part of the agreement, the LNWR were not to use parliamentary powers to acquire any portions of the quay area for their own use. However, they were to be provided with free access onto their lines on the quay for which they would be charged a small annual easement fee. The railway company was also to have permission to construct sidings off the lines on the quay into James Williamson's St George's Works and into Lancaster Gasworks, both of which were situated on the quayside facing the river. The sidings on the quayside were to terminate at the Carlisle railway bridge. However, the agreement made provision for either party to construct, at some point in the future, a railway line to the east of the bridge to make an end-on connection with the line proposed in the bill.

It was expected that the provision of the railway would significantly increase the amount of traffic coming through the port and that this would, in its turn, increase the amount of dock and harbour dues realised. Therefore, in accordance with the Commissioners earlier promise to, in effect, share this increase with the railway company, specific provisions were made for this. At the time of the completion of all the works, an average annual figure of dock and harbour dues would be agreed by reference to the results for the previous ten years. From the date of the opening of the line, any annual excess above this figure would be split equally between the railway and the Commissioners. To keep the dock as competitive as possible, it was agreed that once the line was opened, the Commissioners would not raise the dock and harbour dues without first seeking the agreement of the railway company. In fact, in the years following the opening, the LNWR made repeated attempts to persuade the Commissioners to reduce the dock dues on the cargoes most frequently carried by the railway.

The Act itself received royal assent on 22nd July 1878. As well as authorising the construction of the lines, it conferred parliamentary sanction on the agreement between the LNWR and the Port Commissioners, making it binding on both parties. Indeed, as if to underline this, the full text of the agreement was reproduced as a schedule at the end of the Act. The Act was also slightly unusual in that although it was a railway omnibus Act, promoted by the LNWR, it also conferred upon the Port Commissioners the necessary powers for raising the funds to finance the extension of the river quay at Glasson Dock. This was to take the form of a mortgage on the future dock and harbour dues up to a maximum amount of £30,000. As was usual in these matters, the railway company had three years from the date of royal assent to exercise their powers of compulsory purchase for the land required. They also had five years in which to complete the branch and open it for traffic. The Act of Parliament, incorporating as it did, the agreement of 1878, provided a comprehensive package of interlocking agreements and obligations. On the whole, these worked tolerably well and it was only in the difficult economic climate in the 1920s and 1930s with rail traffic from the dock down to a trickle, that the LMS, as successor to the LNWR, began to fail to honour its part of the bargain.

Chapter Three

Construction and Opening

IN ESSENCE, THE construction of the Glasson Dock branch involved two quite separate but conjoined civil engineering projects. The LNWR would construct the railway, of course, including the separate branch serving St George's Quay at Lancaster. However, as part of their agreement with the railway company, the Port Commissioners had undertaken to extend the existing river quay a further 220 yards upriver to join the railway works of the LNWR. According to contemporary newspaper reports, the extended quay would provide additional berthing space and an extra wharf area of 8,500 square yards. This new facility would also enable larger ships to use Glasson Dock, which might otherwise be prevented from doing so, owing to the limited width of the dock gates.

As we have seen, under the terms of the agreement, the Commissioners had three years from the passing of the Act of Parliament to complete the extensions to the river quay. However, they seemed to have moved rather slowly. It was not until 14th August 1879 that the Commissioners could report that Mr James Mansergh, of London, had been appointed engineer for the pier extension and other new works at Glasson Dock. Even then, owing to delays on the part of Mansergh in preparing the plans and then on the part of the Commissioners who wished to make alterations to them, it was not until 6th April 1880 that the drawings and specifications for the new quay wall were finally signed off. Later that month on the 26th, it was reported that the contract for the extension to the quay wall and the new works at Glasson Dock had been let to Messrs Whitaker and Sons of Horsforth, Leeds, for the sum of £12,790.18s.4d. By early June the *Lancaster Gazette* could report that machinery and plant for the construction of the new quay wall was arriving on site and that men had already commenced work.

On the LNWR side, there appears to have been a similar slowness to get things started. The new line itself presented no significant engineering difficulties, however, the railway company still seemed to be deciding whether to improve the facilities on the canal and, in particular, whether to widen the lock that connected it with the Commissioners' wet dock. There also appears to have been some uncertainty as to how the new railway would interface with the existing canal and dock installations at Glasson. Within its deposit of papers relating to the LNWR, the National Archives holds a handwritten memo, dated 16th July 1879, and headed *Glasson Dock Improvements*:

> *Mr Hull and Mr Slinger to proceed with purchase of land for Commissioners consulting with whoever they appoint to advise with them. Mr Worthington in a like manner to advise with the person they may appoint to carry out their works at the pier. Mr Worthington to make an estimate of the cost of the works utilising the dock only and another estimate for dredging the canal basin and enlarging the lock between the dock and the basin. The Commissioners to be allowed to dredge the canal basin for the purposes of filling up their own works if they elect to do so undertaking to make good any damage if any that may be done to the walls or in other ways.*

In the event, the canal lock was not enlarged. Perhaps the LNWR felt that as the wet dock itself was not going to be altered, there was no pressing need to improve the facilities on the canal. Clearly, it was essential that both the Commissioners and the railway were singing from the same hymn sheet, so to speak, if the project was to be brought to a successful conclusion. It may have taken a little time to achieve this happy condition and this could account for the slowness of both parties to start work. The Commissioners' position was that whilst some enlargement of the wet dock was highly desirable, the cost was quite simply prohibitive. The extension of the river quay, therefore, was seen as the next best thing. The last sentence of the memo requires some clarification. The proposed new river quay wall effectively reclaimed a strip of salt marsh alongside the river and a significant amount of material would be required to backfill the area and bring it up to the level of the existing quays around the dock. The Commissioners had purchased a small dredger in 1870 and so were perfectly capable of extracting material from the bed of the canal basin. In the event, however, gravel for this purpose was obtained from the bed of the river. This, of course, had the added benefit of helping to improve the river navigation for shipping.

An unusual view up the gradient towards the junction with the main line. On the left is part of Williamson's St George's Works situated alongside the main line and fronting onto the Ford Quay alongside the river. Just right of centre is the rear of Lancaster No 4 signal box and beyond it the main line running from left to right.

CRA, Rev J Jackson Collection.

Aldcliffe was the first level crossing at a public road after leaving Lancaster. Here the line crossed open countryside before gaining the estuary of the River Lune, out of sight to the left of the picture. Behind the trees on the right are the distant chimneys of Williamson's Lune Mills.

CRA, Pearsall Collection.

Meanwhile on the railway side of the project, things began to move with the appointment of Messrs Holme & King of Wigan as contractors for the construction of the branch. At the end of August 1879, Mr Worthington, the LNWR's engineer was in discussion with Lancaster Corporation concerning the arrangements for obtaining possession of the portion of the marsh required for the new line and also for the diversion of a water course alongside the railway. Preliminary work appears to have started around this time as the *Lancaster Gazette* for 6th September reported that for the past week, workmen had been engaged in cutting out sods on the land through which the new railway was to pass. By July of the following year, the newspaper could report that the work on the Lancaster–Glasson Dock line was being vigorously pushed on and work had already commenced on the embankments near the River Conder.

In May 1881, the local magistrates and some landowners suggested that the LNWR should appoint a policeman to keep an eye on the men engaged in the construction of the railway. This was immediately agreed to. However, by the following month, it had been arranged that instead of the railway company paying the expenses of the policeman, he was to be engaged and paid for by the contractor. In fact, the local newspapers reveal very few instances of the men working on the construction getting themselves into trouble. In November 1881, John Morris, a man employed on the Glasson Dock railway, was found guilty of being drunk and using obscene language in China Street, Lancaster. In the same month, John Kenny, a navvy, working on the construction of the branch, was found guilty of stealing a one-pint pewter measure from a public house and was sentenced to a month in jail with hard labour. Although coming close together, these do appear to be isolated occurrences. It is possible, however, that there were a number of incidents of what we would today term anti-social behaviour, probably alcohol fuelled, which did not reach the newspapers. As a large town and the home of the forbidding, old county gaol, Lancaster is likely to have had sufficiently strong forces of law and order to keep a few drunken navvies in check. Beyond Lancaster was remote, open, estuarine countryside with no one to annoy except perhaps a handful of local farmers and landowners. As small as it was, Glasson Dock itself was a working seaport with three licenced premises and a constant turnover of British and foreign seamen and casual dock workers. It may be that the drunken high jinks on any ordinary Saturday night would have equalled anything provided by the influx of a few navvies.

The prospect of a new railway line running across the largely empty expanse of Lancaster marsh, on the western edge of the town, stimulated some interest in developing new industrial sites on virgin land, which could be conveniently served by the railway. There had been some industrial development in this area as early as 1863, when the newly formed Lune Shipbuilding Co. rented an area of the marsh from Lancaster Corporation. This fronted onto the River Lune opposite the New Quay. The company subsequently purchased the land outright and until going into voluntary liquidation in 1870, constructed a succession of well-regarded, large, wooden sailing vessels. Once launched, these were fitted out alongside the New Quay. In May 1871, the shipyard site was acquired by James Williamson who was a manufacturer of oil cloth, producing table and wall coverings, shelf borders and window blinds. The basis for these products was cotton cloth, woven in Williamson's own mills on the east side of the town and then carried across to his works on St George's Quay overlooking the river. Here it was coated with a mixture known in the trade as composition or kivver, which included amongst other ingredients, china clay, chalk, resin and linseed oil. During the course of manufacture, the oil cloth could be cut, embossed, varnished, painted, or printed to produce a wide variety of items attractive to the Victorian homemaker. The firm quickly diversified into the production of floorcloth, in essence, a more durable version of oilcloth and, consequently, tended to use hessian as a backing material instead of cotton. The activities of Williamsons were subsequently to have a significant impact on the marsh, on the traffic passing up and down the Glasson Dock branch, and, indeed on the town of Lancaster itself.

It is appropriate at this point to provide some details of the firm's history. Around 1855, James Williamson Senior built a new factory on a site fronting onto St George's Quay and immediately to the west of the Lancaster and Carlisle railway line. This became known as St George's Works. In 1862, as the business expanded, the works was extended westwards along the quayside. Having obtained possession of the old shipyard site, a little further down the river, James Williamson, now in business with his two sons, Thomas and James Junior, began to develop this for the manufacture of floor cloth, whilst at the same time retaining the existing works fronting onto St George's Quay. This new site quickly became known as Lune Mills to distinguish it from the older St George's Works, a little distance further up the river. Williamson's therefore had a presence on Lancaster marsh well before the passing of the 1878 Act of Parliament authorising the construction of the Glasson Dock branch. Shortly after the death of his father in 1879, James Williamson Junior purchased his brother's share in the business and assumed sole control — a control he was not to relinquish until his death in 1930. The eventual course of the Glasson Dock branch sliced obliquely across the undeveloped southern end of the Lune Mills site, so it would have been a simple matter to construct a siding alongside the main running line and into the works itself. It was certainly intended that the site should be so connected. The *Lancaster Gazette* for 20th May 1882 reported that after several meetings and the submission of an amended plan, the Corporation had agreed to the sale of a strip of land to Williamson for this

Glasson station. The driver looks on as the guard and the shunter leave the brake van and walk down to the front of the train to supervise the shunt onto the Commissioners' sidings alongside the quay. Once the two men are in position, the locomotive will slowly push the wagons onto the quayside and position them alongside the vessel being loaded or unloaded. Taken in March 1962. Ron Herbert.

purpose. James Williamson had also approached the LNWR some twelve months earlier as a Special Committee minute, dated 20th May 1881, recorded that plans had been approved for siding connections from the branch to the Lune Works, subject to agreeing suitable terms with the firm. However, for reasons which are unclear, the sale of the strip of land was not proceeded with and the private siding was not constructed. Instead, once the line was opened, Williamson's appear to have made extensive use of a long siding running at the rear of the Lune Mills site. This siding, the property of the LNWR, was in effect a rather long extension of the head shunt for the line running onto St George's Quay.

Meanwhile, construction of the railway to Glasson Dock was proceeding at a pace. In August 1881, an LNWR Permanent Way Committee minute noted acceptance of a tender for three cottages to be built on the Glasson Dock branch. These were for the accommodation of the crossing keepers at the manned level crossings at Aldcliffe, Stodday and Conder Green. Plans for these do not appear to have survived. However, evidence from maps and photographs suggests that, whilst they were constructed to a similar design, they were not identical. There is also evidence suggesting that the cottage at Conder Green was not constructed until shortly after the line was opened.

Although the arrangements concerning the construction of the railway and the allocation of the dock and harbour dues had been agreed between the railway company and the Commissioners, there was still some fine detail to be resolved. At Lancaster, where the branch to St George's Quay ran onto the Commissioners' property alongside the Lune, it was necessary to agree a sum for the right of easement. This was fixed at £20 *per annum* in perpetuity and was subject to the existing easements and access to the quay granted to Williamson's and also to the Lancaster Gas Co, both of which had premises facing the river. Further discussions also took place in connection with the railway sidings for both these concerns.

At Glasson Dock, things were slightly more complicated. The LNWR already had possession of a significant stretch of land alongside the canal basin by virtue of its lease of the Lancaster Canal. However, in addition to this, the railway acquired an adjoining strip of land alongside the road into the village between Brows Farm and the Victoria Hotel. This was purchased from the Commissioners for £250. There was some discussion about the railway acquiring the whole block of buildings adjacent to the dock, including the Victoria and Pier Hall Hotels. In the event only a row of seven cottages at the rear of the latter establishment was acquired at a cost of £900.

During April 1882, the Commissioners were exercising their minds in connection with the arrangement and specification of the lines of rails to be laid on the quays at Glasson Dock. Mr Mansergh, the Commissioners' engineer, was instructed to communicate with Mr Worthington, the LNWR's engineer, in connection with these matters. Mansergh was also requested to avoid turntables on the quayside if possible, although in the event, one was installed on the pier head. It had been decided that the Commissioners' lines on the dockside would be worked by steam cranes and this was to have a bearing on the positioning of the lines. After some discussion, it was agreed that the lines used by the cranes should be as close to the edge of the quay as was practicable. Two of the Commissioners were appointed to visit Fleetwood and Garston docks to obtain further information as to the distance between the edge of the quay and the line of rails of the crane road and also the distance between the crane road and the inner road upon which the wagons would stand. All this indicates that the method of operation was for the crane to operate up and down the track closest to the quay side with the wagons being loaded or unloaded, standing on the inner adjacent road. This is certainly borne out by a number of photographs of the dock in the late nineteenth and early twentieth centuries. However, in spite of all this discussion, it was not until early 1883 that the plan of the layout of the lines on the dockside was agreed upon. An LNWR Special Committee minute of 19th January 1883 provides the following information:

Submitted Minutes of Meeting between Messrs Cattle, Slinger and Gill (for Mr Worthington) & the Lune Commissioners together with approved plan of the lines proposed to be laid down upon the quay at Glasson Dock, and

Resolved — That upon the plan being signed by the Chairman of the Lune Commissioners and upon formal

A view looking towards Glasson Dock across the curved Conder viaduct. This shows to good effect the construction of the decking and the timber baulks and associated transoms upon which the rails are fixed. A checkrail is fitted to the inside rail. The signal at the far end of the viaduct is the distant for Glasson Dock.

CRA, Rev J Jackson Collection.

application being made by them for the work to be carried out on the terms of the agreement scheduled in the Act of 1878, Mr Worthington be authorised to proceed with the works at the cost of the Commissioners.

Mr Cattle, or Henry Cattle to give him his full name, was the District Traffic Superintendent for the Lancaster and Carlisle Division of the LNWR and based in Lancaster. From this point on, he was to be the first point of contact for the Commissioners in relation to all railway matters. In the event, both the river quay and the wet dock were provided with their own set of double-track sidings. The line to the wet dock was an extension of the loop in front of the passenger platform at the station. After crossing the road into the village, this ran alongside the canal basin thus providing rail facilities for any vessel moored there. Access to the new river quay was also from the run-around loop at the station, with the line running alongside the estuary and onto the newly constructed river quay.

Whilst the arrangement of the facilities at Glasson Dock was being fine-tuned, discussions were also underway to finalise the plans for the railway infrastructure on St George's Quay at Lancaster. In 1880, the privately owned gasworks on the quay had been taken over by Lancaster Corporation and in June 1882, the Commissioners met with the LNWR and the Corporation Gas Committee to discuss the laying of rails onto St George's Quay and into the gasworks. This had already been agreed in principle by all the parties concerned and it was simply a question of arranging the final details. An LNWR Special Committee, minute of 16th June 1882, provides the following information:

Resolved — That the work be carried out in accordance with this plan at the cost of the Corporation, a turntable being substituted for the curve shown on the plan if found to be more convenient, subject to the Commissioners agreeing that in addition to the single line of rails already laid down upon the St George's Quay arrangements be made for laying down a second or loop line for accommodation of empty wagons and for performing the loading and unloading of Goods Traffic so far as may be necessary upon the Quay.

That the siding of Messrs Williamson & Sons works shown in pencil on plan be laid in at their cost.

Once closure of the Glasson Dock–Freeman's Wood section of the branch was announced, a final rail tour was hurriedly organised by the Railway Correspondence and Travel Society. Consisting of only the locomotive and a string of goods brake vans, the train is seen here coming off the branch and into the station at Lancaster Castle. The engine appears to have had some attention from the cleaners and the numberplate has been painted blue and picked out in white. The date is 20th June 1964.

Derrick Codling.

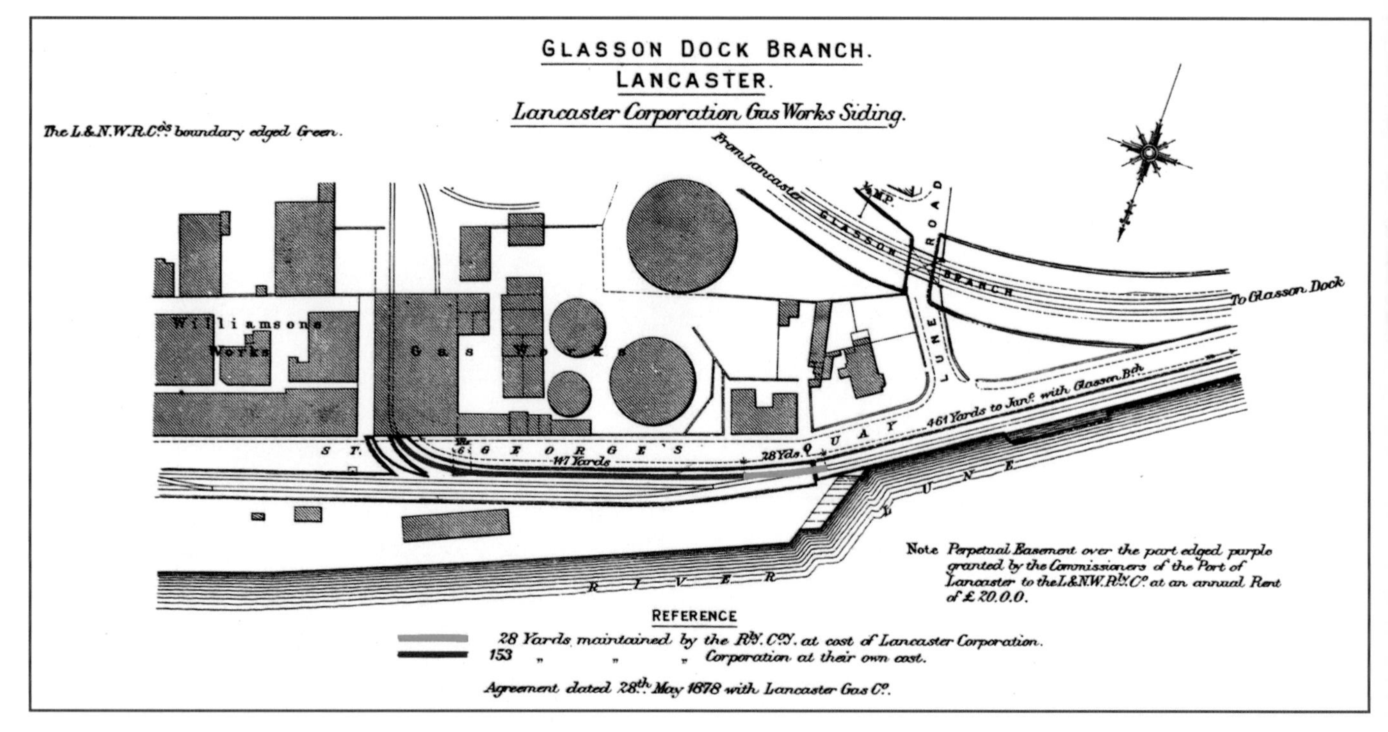

The LNWR siding diagram showing the arrangements for the private siding for the gasworks. Of interest is the easement provided for the proposed private siding to serve Williamson's St George's Works. In the event, this was never constructed. Also shown is a crossover connecting the two sidings on the quay. This appears on some LNWR plans but is absent from all the editions of the 25-inch Ordnance Survey.

CRA, Peter Robinson Collection.

In the event, a turntable was not installed and instead the siding curved sharply across the road to enter the gasworks. The LNWR diagram relating to the private siding gives the date of the original agreement as 28th May 1878. This, therefore, predates the Act authorising the construction of the railway. The second line extending along the quay was agreed to and subsequently put in place. However, it is difficult to decide whether the two lines were ever connected at the Carlisle bridge end, so as to enable engines to run around their trains. There are two LNWR plans showing the lines on St George's Quay, but these are unfortunately undated. One shows a cross-over connecting the two sidings and the other does not. Certainly, the various editions of the 25-inch Ordnance Survey do not show it. Clearly, plans were also being made for a private siding to serve Williamson's St George's Works and it appears that this was to be in addition to the two sidings on the quay side. Indeed, the plans referred to above, which show the extent of the LNWR's right of easement on the quayside, make provision for a line running off the quay and into the street separating St George's Works from the gasworks. However, all the available evidence indicates that this was never built. Various editions of the Railway Clearing House *Handbook of Stations etc* contain an entry for a Williamson's Siding, which was situated on the Lancaster Quay sidings. This seems to suggest that the firm enjoyed some sort of private siding facility here. However, the absence of any private siding agreement or diagram makes it impossible to say exactly what the relationship between the firm and the railway company was. What can be said with certainty is that for many years, Williamson's made extensive use of the two sidings on the quay and, indeed, of the single line running along the rear of the Lune Mills site. All of this railway related work was carried out by the LNWR, including the continuation of the private siding inside the gasworks, for which the Corporation paid an agreed sum. Extracts from Lancaster Corporation minutes indicate that the estimated cost was £350. The LNWR committee minute above confirms that there was a single line of rails on St George's Quay as early as June 1882. However, the work on the siding serving the gasworks does not appear to have commenced until after January 1883.

Meanwhile, at the Glasson Dock end of the branch, work was progressing on the new river quay and on the railway itself. By the beginning of October 1882, the Commissioners' contractors Whitaker Bros, had completed the former and a cheque was prepared, which constituted full and final payment for the work carried out. The LNWR had decided that there would be a small passenger station at Glasson and that this would be sited overlooking the estuary, just a short distance outside the village. In January 1883, a timber station building was ordered from WL Cleminson of Lancaster at a cost of £520. A small station master's house was also provided and this was situated opposite to where the line to the wet dock was to cross the road into the village. However, it was not constructed by the LNWR. Rather, it was a pre-existing building, constructed of local stone, originally called Parsonage Cottage, and dating from the 1850s. Presumably the house had been lying empty and the railway company acquired it at the time the line was built. There was some discussion about the provision of a small goods shed for the station. However, it was felt that the level of local traffic would not justify the expense, and therefore the facilities were limited to a short siding serving a small platform with an end loading dock.

The LNWR was especially keen that the dock and harbour dues at Glasson should be as competitive as possible so that existing traffics could be lured away from other ports in the area. The docks at Barrow were the property of the Furness Railway and the pier at Morecambe was served exclusively by the Midland. The LNWR had a one-third stake in the port of Fleetwood, by virtue of its interest in the Preston & Wyre lines. However, the remaining two-thirds were held by the Lancashire and Yorkshire Railway. Glasson Dock, therefore, was the only port on Morecambe Bay to which the LNWR had exclusive access. In any case, having invested in a railway to the port, it was imperative that the capital expended began to show a satisfactory return for the LNWR shareholders. Under the terms of the 1878 agreement, the LNWR was to have an equal share in any increase in the dock and harbour dues. Any reduction in the rates charged by the Commissioners, therefore, would affect the railway company up to a point. However, for the LNWR, the real profits would come from carrying the railway traffic to and from the port. The Port Commission was what we would today call a non-profit making organisation, with all surpluses of income over expenditure being ploughed back into maintaining the river navigation and the facilities of the port as a whole. The dock and harbour dues were a principal source of income for the port, so there was always some reluctance on the part of the Commissioners to reduce these.

The Commissioners minutes show that, in the early part of 1883, even before the line was opened, the LNWR had pressed them to reduce their dock dues on certain commodities. In particular, the railway company seems to have set its sights on

diverting pig iron and iron ore traffic into the port. These were of course heavy, bulky commodities, which, realistically, could only be carried away by rail. For their part, the Commissioners had discussed the matter amongst themselves at some length and had compared their own charges with those at Barrow, Morecambe and Fleetwood. Following this, they did agree a reduction in the rates for vessels carrying these commodities. However, whilst the Commissioners wished to meet the requests of the railway company, they would only go so far and a request a little later on, for further reductions was met with a polite refusal. This whole question of the dock dues was a subject to which the LNWR would return time and time again once the line was open for traffic.

By January 1883, the new lines and signalling arrangements at the junction for the branch had been completed and were ready for inspection by the Board of Trade. This was duly carried out on the 27th of that month. Perhaps in anticipation of heavy traffic from both the Glasson Dock and the St George's Quay line, the LNWR had provided a double track junction and the two lines ran parallel down the branch for just under a mile. However, away from the junction with the main line, these were not physically connected, and were operationally separate. The idea being that traffic could move to and from the lines serving St George's Quay and the rear of the Lune Mills site without interfering with traffic on the main running line to Glasson Dock. This separate but parallel section of the Glasson Dock branch, including the line down to St George's Quay itself, is generally referred to in LNWR documents as the Lancaster Quay goods line. For the sake of clarity, this convention will be adhered to throughout this narrative.

By March 1883, the line itself appears to have been completed although the station at Glasson Dock was still unfinished and there remained some work to do on the sidings belonging to the Commissioners. The *Lancaster Gazette* for 18th March 1883 reported that, on the previous Wednesday (the 14th), some of the Port Commissioners had made their first trip down the line to Glasson Dock, accompanied by Mr Cattle, the LNWR's District Traffic Superintendent, and other representatives from the railway company. The railway had placed a saloon carriage and locomotive at the disposal of the party and these were decked out with banners to mark the occasion. Following an inspection of the new works, the party dined at the Pier Hall Hotel at Glasson Dock. The newspaper article explained that the first ship to sail from Glasson Dock with a cargo brought in by rail was likely to be the *Clyde*, currently undergoing repairs in the graving dock. This was one of the vessels engaged in the Canadian timber trade. She would be ready to

2MT No 46422 pauses during operations at Glasson Dock. The gentleman in the mackintosh looking into the cab is the travelling shunter who has accompanied the engine from Lancaster. His task was to open and close the level crossing gates and then to control the train movements at the terminus. Ron Herbert.

sail in about three weeks time and, prior to departure for Quebec, would be loaded with 500 tons of coal brought straight from the pit in railway wagons. At this point, there were no mechanical handling appliances at Glasson Dock, so the coal would have had to have been transferred into the ship's hold using only manpower and shovels. However, all of this suggests that the line was being used for goods traffic as early as April 1883.

In March 1883, with the completion of the railway now on the horizon, the Commissioners turned their thoughts to the provision of steam cranes for use at Glasson Dock. Two of their number were tasked with looking into the matter and, in April, they reported back with specifications from Messrs Appleby Bros of London and JH Wilson & Co of Sandhills Works, Liverpool. After careful consideration, the two gentlemen were authorised to purchase two steam cranes on the best possible terms, but it was stipulated that they should also confer with Mr Cattle of the LNWR before finalising any purchase. From the outset, the Commissioners had made it clear to the railway company that they did not wish to play any part in the loading or unloading of vessels. It does appear that this was the usual practice at Glasson Dock, with the recipient of any cargo arranging for casual labour to discharge the ship. Therefore, although the two steam cranes were to be the property of the Commissioners, it was agreed that they would be operated exclusively by the LNWR. Probably at Mr Cattle's behest, advice was sought from Francis Webb, the LNWR's Chief Mechanical Engineer. Webb recommended a firm in Birkenhead by the name of Taylor and two of the Commissioners duly visited their works and saw one of their cranes in action. However, following a visit to Lancaster by Mr Appleby, one of the principals of Appleby Bros, it was decided to order two of their cranes. Delivery was to be on the LNWR in London, and both cranes were to be erected, tested and ready for work for £450. The relevant Commissioners' minute contains a transcript of the letter of offer from Messrs Appleby and provides a detailed specification for the cranes. The boiler and all the machinery, apart from the jib, were to be hidden inside a substantial wooden housing. They were capable of moving under their own power and there was to be a steam

Glasson, captured in repose between trains. This was taken in the early years of the twentieth century and there is a slight sense of untidiness and clutter in the scene, denoting a working railway station in its everyday guise. The level crossing is still protected by a double signal with one arm for each direction.

CRA, LCR Collection.

winch on each crane, capable of hauling wagons. Following advice from Mr Cattle, the Commissioners also ordered four iron buckets to go with the cranes. These were to be used for loading/unloading loose cargoes, although their principal use initially, was for the discharge of iron ore. The buckets were provided by Alsop Bros of Preston.

With the construction of the branch now nearing completion, the LNWR wrote to the Board of Trade on the 11th May 1883 to inform them that it was intended to open the Glasson Dock branch for the conveyance of passengers. This was followed up on the 27th of June with a note saying that the line would be ready for the safe conveyance of passengers by the 29th of that month and would be ready for inspection at any time during the ten days following the 29th. This sequence of rather formal letters was a prelude to the necessary business of getting the Board of Trade to provide a certificate to the effect that the new line was suitable for the safe carriage of passengers. In due course, Colonel Rich, one of the Board's inspectors came up from London and went over the line, accompanied by Mr Worthington, the LNWR's engineer; Mr Thompson, principal of the signalling department, Crewe; Mr Stockdale, superintendent of the telegraph department; and Mr Cattle, the District Traffic Superintendent. The Colonel's report dated 7th July 1883 survives in the Board of Trade papers held at the National Archive. The first thing to say is that the inspection confined itself to the parts of the line that were to be used by passengers. Therefore, there is no mention in the report of the Lancaster Quay goods line, which was intended to be used for goods traffic only. In fact, as we have already seen, there is evidence that there was at least some goods traffic running between Lancaster and Glasson Dock perhaps as early as April of that year. In his report, the Colonel described the line as 5 miles and 44 chains long with the steepest gradient being 1 in 50 and the sharpest curve having a radius of 9 chains. He commented that check rails had been fixed around all the sharp curves. The permanent way consisted of bull head rails having a weight of 75lb per lineal yard. The rails were joined with fish plates and fixed with wooden keys into cast iron chairs weighing 30lb each. These chairs were fixed to transverse sleepers with one spike and two corrugated iron screws. The sleepers themselves were laid, on average, three feet apart and were nine feet long, ten inches wide and five inches deep.

There was one brick-arched overbridge and eight underbridges. Of these, one was constructed entirely of brick, one had wrought iron girders on brick abutments, and the remaining six had cast iron girders on stone or brick abutments. The curved viaduct over the River Conder consisted of three openings of 25 feet and two of 30 feet. The abutments and piers were constructed of stone and brick and the girders of wrought iron. The Colonel commented that the works appeared to be substantially constructed. There was an issue with the gates on the level crossing at Stodday, which did not close across both the road and the railway. He requested that this be rectified and that a lamp and a crossbar signal be provided to protect the crossing. He also noted that the station building at Glasson Dock was in an unfinished state. Colonel Rich had apparently asked some searching questions relating to the building's construction, as following the inspection, William Worthington, the LNWR's engineer, provided the Colonel with a tracing showing the arrangement of the cast iron brackets supporting the canopy over the platform. Worthington also assured the Colonel that the incomplete building would be fenced off from the platform until construction was completed, a temporary ticket office being provided in the meantime. Notwithstanding these minor issues the Colonel professed himself satisfied that the line was ready to be opened to the public. He would issue a certificate to that effect, upon receipt of an undertaking as to the method of train control to be adopted for the part of the branch between Castle station and Glasson Dock. The LNWR wrote to the Board of Trade on the 11th July, confirming that the trains on this section of the branch would be controlled using the absolute block telegraph and the train staff systems combined, and enclosing a signed and sealed undertaking to that effect. Upon receipt of this document, it was simply a matter of the Board of Trade issuing the formal certificate and the paperwork was completed. In fact, the LNWR had jumped the gun slightly and opened the line for passengers on the 9th, but no one seemed to mind.

The *Lancaster Gazette* for Saturday 14th July provided extensive coverage of the opening on the previous Monday, together with some illuminating details relating to the state of trade at Glasson Dock prior to the coming of the railway. In the village, there were no formal celebrations marking the opening of the line for passengers. However, a crowd of people gathered to witness the departure of the first train to Lancaster and the engine itself carried banners in honour of the occasion. A timber ship unloading in the dock was decorated with flags and there was some evidence of bunting at the shipyard and in one or two other places. Fifteen passengers were booked for the first train from Glasson and between 30 and 40 on the first train from Lancaster at 9-15. The newspaper commented that for many years past, the principal commodities coming into Glasson Dock had been timber and grain. However, of late, the latter trade had fallen off owing to the limited width of the dock gates and the gradual increase in the size of the ships carrying this commodity. At the time of the opening of the railway, the vessel *Patriot Queen* was in the wet dock discharging a cargo of timber for the Preston Timber Co. This was loaded into barges and continued its journey to Preston *via* the Lancaster Canal. Timber, principally from Canada and Scandinavia, was a very well-established traffic through the dock and in addition to the Preston firm, two Lancaster timber merchants regularly received cargoes at the port. Contemporary reports indicate that at this time, a portion of the canal basin was used as a timber pond for the storage of imported logs. It would appear that once the branch was opened, much of this traffic left Glasson Dock by rail. Certainly, photographs taken around the dock in the late nineteenth and early twentieth centuries show stacks of sawn planks piled high on the quayside and in some instances, railway wagons loaded with timber.

Chapter Four

Consolidation

ALTHOUGH THE BRANCH was now opened for goods and passenger traffic, there were still several matters that required attention to make the best use of the extended port facilities.

It had been decided, perhaps as early as September 1882, that some sort of shed or warehouse should be provided at the dock in which to store cargoes under cover. However, it was not until October 1883 that the plans and specifications for it were finalised. It had originally been intended to site this on the dockside close to the berth nearest to the outer dock gates. However, in the event, it was erected much closer to the passenger station, alongside the canal basin. No good photograph of the shed in its original condition has come to light but it was a large structure divided into two bays, each with its own separate curved roof. Its purpose was unmistakable, as on each end, emblazoned in large letters was the legend *Storage Sheds*. All the references to the structure in the Commissioners' minutes indicate that it was to be built by the LNWR and that the cost would be borne by the Commissioners. In its originally proposed location next to the wet dock, it would have stood on the Commissioners' land and, therefore, would have become a part of the dock estate. However, the change of location onto what was canal company land, seems to have prompted a decision that the shed would be both constructed and owned by the railway company. Certainly, an LNWR legal map, dating from the early 1890s, indicates that the sheds were the property of the Euston company.

Shortly after the opening of the line, the LNWR received a request for the provision of a private station from the owner of Ashton Hall, Mr JPC Starkie, who also happened to be the MP for the North East Lancashire constituency. Part of the route of the branch ran through Mr Starkie's land, and having accommodated the railway company, perhaps he felt that it was appropriate to ask for a favour in return. The LNWR Special Committee Minute of 17th August 1883 provides the following information:

> *Mr Starkie's application for trains to be stopped at a place near his house on the Glasson Branch for the convenience of himself and Family, was agreed subject to Mr Starkie paying the costs of creating a proper platform &c. the Company reserving to themselves the right of discontinuing the privilege at any time if they find it necessary to do so, and on the understanding that Mr Starkie indemnifies the Company from all risk of accident.*

The LNWR provided a short timber platform and wooden shelter, overlooking the estuary and situated in front of a cottage on the Ashton Estate known as *Nan Bucks*. Railway documents from the nineteenth century refer to the station as Ashton Platform, but by LMS days, it was known as Ashton Hall Halt.

At the beginning of October 1883, with the delivery of the steam cranes imminent, thoughts turned for the question of the supply of water for their boilers. It should be explained that at this time and for many years afterwards, there was no mains water supply at Glasson Dock. The villagers obtained their water for domestic purposes from shallow wells or collected rainwater in stone cisterns. As far as the LNWR was concerned, this did not create any difficulty. At the company's engine shed at Lancaster, the water supply for the engines came from the Lancaster

The platform at Ashton Hall Halt. This was taken around 1948, some 18 years after the closure of the line to passengers and the timber platform edge is showing significant signs of decay. The house in the background is *Nan Bucks*, which was part of the Ashton Hall Estate.

CRA, Rev J Jackson Collection.

An unidentified sailing vessel at Glasson Dock around 1900. The railway wagons on the left appear to belong to the LNWR. The load is stacked high in the wagons, suggesting that it is cork wood for Williamson's or wood pulp for Cropper's of Burneside. The building on the far right is the grain warehouse, which dated from the opening of the canal to Glasson, and was demolished in 1939. Courtesy Lancaster City Museums.

Canal. It was, after all, the railway company's own water and, therefore, would have cost nothing. However, when the LNWR chemists came to test the water in the basin at Glasson Dock, they found that it was unsuitable for use in steam boilers, owing to the relatively high salt content. From the Commissioners' perspective, there was another matter to consider. Ships arriving at Glasson often needed to take on water for the crew's use and for many years this had, almost unbelievably, been obtained from the canal basin. The Commissioners, not surprisingly, were unhappy with this state of affairs. Aside from the obvious health concerns, it was felt that it would perhaps make the port a little more attractive, if a pure water supply was available. In the event, the LNWR solved the problem as far as the cranes were concerned, by piping the water from one of the canal locks above Thurnam and transferring it into a tank at Glasson. This ensured that the water was free from contamination with sea water and the difference in height provided sufficient pressure for an adequate supply at the dock side. This still left the question of the supply of water to vessels at Glasson Dock and, whilst the Commissioners and the various arms of local government agreed that it was desirable that both the village and the vessels in the dock, should have a supply of pure drinking water, very little was done to achieve this end. In October 1899, with no further progress having been made, the Commissioners

Ashton Platform or Ashton Hall Halt as it was known in LMS days. This plan is derived from an LNWR legal map dating from the early 1890s, and the latter shows the entire branch from Lancaster Castle Station to the pier head at Glasson Dock. Map redrawn by Alan Johnstone.

ASHTON PLATFORM
Level Crossing
to Lancaster
NAN BUCKS
PLAT FORM
to Glasson Dock
12 inch Pipe
12 inch Pipe

A wonderful evocation of Glasson on a Summer's day. An unidentified Coal Tank simmers at the platform while the driver, fireman and guard, together with the entire station staff face the camera. Any passengers from the train have dispersed and the locomotive has yet to run around its two-carriage train in readiness for the return journey. An attempt has been made to brighten up the platform by creating flower beds.

Commercial postcard. Author's Collection.

resolved that they would supply water to the vessels at Glasson Dock themselves on a temporary basis. They approached the LNWR to ascertaining the cost providing a railway tank wagon and ferrying it between Lancaster and Glasson Dock when it required filling with water. However, the costs were deemed to be prohibitive, and the matter was once again allowed to drop. It was not until 1933 that the village finally received a mains water supply from the Fylde Water Board; the costs of connection being met by the Port Commissioners. It was only at this point that the problem of supplying water for vessels in the dock was finally resolved.

It was not until mid-October 1883, after the opening of the branch, that the two steam cranes finally arrived at Glasson Dock and on the 18th of that month, the Commissioners met on the dockside with representatives of the LNWR and Messrs Appleby to watch the cranes being put through their paces. There were one or two minor additions and alterations requested by the men who would be working the cranes and these were readily agreed to by the man from Appleby's. It was noted that slings would be required for handling the pig iron traffic, which the LNWR was expecting to obtain. The Commissioners decided to order twelve of these at once, together with chains. They were to be copies of the ones in use at nearby Fleetwood. From the outset, it had been decided that the cranes would be operated by the LNWR, however, in accordance with the 1878 agreement, the railway also agreed that it would take responsibility for maintaining them. A list of spare parts provided by Appleby's, was passed to Mr Webb, the railway company's Chief Mechanical Engineer at Crewe, who felt that it would be more efficient if his company kept a supply of spare parts for use when required. The railway company also supplied the coal for the cranes and a close scrutiny of a particular photograph taken at Glasson Dock around the turn of the century, reveals an LNWR locomotive coal wagon standing on the quayside. Presumably a loaded wagon was dispatched down the branch from Lancaster shed as and when required. A different photograph from around the same period shows an old LNWR Wolverton locomotive tender standing alongside the dock. It's precise purpose in unknown. However, it is possible that it was used to supply both coal and water to the two cranes.

The two steam cranes were to become so much a part of the scene at Glasson Dock and at least one of them can be spotted in most of the photographs taken there in the late nineteenth and early twentieth centuries. They did not have an entirely uneventful existence however. In November 1888, a vessel was being moved to a loading berth on the east side of the dock when a part of it, probably the bowsprit, came into contact with one of the cranes, which was standing idle on the quayside. The collision caused considerable damage to the wooden housing over the boiler and machinery. Captain Greenwood, the harbour master, was instructed that, when the cranes were not in use, he must give notice to the railway officials at Glasson to have them moved to a place of safety.

In November 1883, it must have seemed that the investment in the facilities at Glasson Dock was beginning to bear fruit. On the 12th of that month, several of the Commissioners assembled on the quayside to watch the first cargo of iron ore being discharged straight into railway wagons and taken out along the new branch line. The Commissioners' records don't reveal the destination of this first cargo, but it is very likely that the ore was being sent to the ironworks at Carnforth, just a few miles to the north of Lancaster. The transfer from the ship's hold to the wagons was accomplished using the new steam cranes fitted with the buckets.

With traffic in the dock beginning to show signs of increase, the LNWR turned its attention to providing better facilities for shipping coal from Glasson Dock. The Euston company's lines served much of the nearby Wigan coalfield, consequently the coal could be brought directly from the colliery to the quayside, thereby providing an additional source of traffic for the branch. Early in 1884, Mr Cattle wrote to the Commissioners requesting permission to place a coal tip on the river quay at Glasson. As usual, the Commissioners decided to form a small committee to look into the matter. However, in August 1884, a group of the Commissioners convened at Glasson to meet with representatives of the LNWR, including Richard Moon, the chairman, to discuss the wider question of increasing the accommodation at the port. Clearly Moon felt that it was time to visit the company's new branch line and review progress so far. According to the Commissioners' minutes, Moon was not averse to some improvement to the existing facilities, but given the depressed state of trade at that particular time, he considered that it would be advisable to defer any expenditure for the moment. However, in this instance, the Commissioners would be footing the bill, so in November of that year a group of them met with Mr Cattle and Mr Adamson of the LNWR, to decide where best to site the coal tip on the river quay. As the appliance was to be installed on land owned by the Commissioners, it would become part of the dock estate. However, as with the steam cranes, it would be operated and maintained by the railway company. The estimated cost was to be about £450, and the LNWR agreed to construct and complete the tip and charge the Port Commissioners the cost price only. No work appears to have been carried out on this until early 1886. In February of that year, Mr Cattle wrote to the Commissioners requesting permission to erect a railway weighing machine and hut on the Commissioners' land close to the row of cottages owned by the LNWR. This was agreed to at an estimated cost of £140, again the work being carried out by the LNWR and charged to the Commissioners at cost. By April of 1886, both the coal tip and the weighing machine were almost complete, and the Commissioners made arrangements to have the river berth at the tip levelled prior to bringing it in to use. In passing, it should be noted that whilst vessels in the wet dock at Glasson remained afloat throughout their stay, those berthed

This is an enlargement of a commercial postcard showing the area around the pier head at Glasson Dock around the turn of the century. On the left is the old watch house and next to it the ex-LNWR slotted semaphore signal, used for communicating with the river pilots at Sunderland Point. The travelling crane is not one of the two cranes belonging to the Commissioners and has presumably been brought in by the LNWR, either as a temporary replacement, or as a third appliance to handle the traffic at the busiest times. On the far right are stacks of sawn timber, so much a feature of the quayside for many years.
Courtesy Lancaster City Museums.

at the new river quay or indeed, at the quays further upstream at Lancaster, were subject to the fluctuations of the tide. When this receded, the vessel settled onto the riverbed until re-floated at the next tide. This was perfectly acceptable providing the riverbed was level and even. Where it was not, the weight of the vessel, especially if loaded with cargo, would be distributed unevenly across the bottom of the hull and damage or strain might result.

The tip was brought into use probably in September 1886. However, in October 1887, the Commissioners received a letter from Mr Cattle enclosing a note from S Thompson & Co, a prominent local firm of coal factors, saying that they would not load any more vessels at Glasson Dock unless some alteration had been made to the berth under the coal tip. The precise nature of the problem was not revealed, but it was probably that the river bed at that point had become uneven. This was a regular problem with the berths further upriver at Lancaster and seems to have been caused by freshwater floods bringing sand and gravel from further upstream. The Commissioners replied to Mr Cattle, assuring him that the berth under the tip would be attended to. Notwithstanding the problems with the berth, in the year ended 19th September 1887, 6,026 tons of coal were shipped at Glasson Dock. At this time, most of the coal wagons owned by the collieries and coal merchants had a maximum capacity of around eight tons. Dividing 6,026 by eight gives 753.25 so that is equivalent to something like 800 wagons over the year, travelling down the branch to Glasson Dock, with the same number of empties returning to the main line. In the following year, the total tonnage shipped *via* the coal tip was 6,598 showing an increase of 572 tons over the previous year's figure. Alas, after 1888, apart from a flurry of activity towards the end of 1893, the coal exports appear to have become infrequent and the tip must have seen little use. In any case, by the early years of the twentieth century, it seems to have become the practice to use the cranes and buckets whenever coal needed to be loaded into vessels. Although the shipping traffic had almost dwindled away, coal continued to be brought regularly to Glasson Dock for bunkering steam ships and photographs exist showing wagons from a number of different collieries standing on the quayside sidings.

Another item of equipment that was installed at Glasson Dock around this time was, perhaps, a little more unusual. Both the channel to the wet dock and the river channel up to the quays at Lancaster could be difficult to navigate, and consequently, the Commissioners employed a number of river pilots to ensure that vessels reached their destinations safely. Some of these lived away on the other side of the Lune estuary at Sunderland Point. There were no telephones in this remote spot and it was often difficult to attract the attention of the men on the other side to inform them that their services were required. Someone, probably the harbour master, Captain Greenwood, hit on the bright idea of using a railway type semaphore signal to pass instructions to the pilots at Sunderland Point. In October 1884, some enquiries were made of Mr Cattle of the LNWR and he obligingly wrote back to say that the signal department of the railway company could supply a second-hand semaphore signal with a height of 20 feet, for about £10. This was accepted and the Commissioners' clerk was instructed to confer with Captain Greenwood as to the best mode of signalling and to report at the next meeting. The signal was erected next to the watch house on the dock pier head, where it could be most easily seen from across the river. There is at least one clear photograph showing the whole signal. There is also a charming sketch of it, in the relevant Commissioners' minute book. It was a slotted signal, that is to say, when not in use, the arms disappeared inside the post. From this point onward however, the method of operation differed from that encountered on the railway. There were three arms and each could be separately raised to one of three different positions using a dedicated lever near the base of the post. An arm raised upwards indicated that a pilot was required at Lancaster. An arm raised to the horizontal position showed that a pilot was required at Glasson Dock. Finally, an arm pointing downwards indicated that the pilots were to look out for a vessel coming in on this, or the next tide. By arranging the signal in this manner, all three messages could be sent simultaneously should that be required. The structure appears to have survived more or less intact until blown down in a gale early in 1927.

Stodday was the intermediate level crossing on the branch and served a public road running from the village to the salt marsh on the estuary. At all of the three level crossings, a two-story cottage was provided for the crossing keeper. However, these posts were abolished in 1931 and, instead, the gates were opened by a travelling shunter, who accompanied the train to Glasson Dock.

CRA, Rev J Jackson Collection.

Whilst the facilities at Glasson Dock were being fine-tuned, the branch onto the quay at Lancaster was not forgotten. In May 1883, the Commissioners received a letter from Mr Cattle, requesting permission to erect a small office on St George's Quay near to their sidings. This was granted, subject to an annual rent of five shillings. The wooden hut, with a board running along the apex of the roof bearing the legend *L & NWR Inquiry Office* can be clearly seen in at least one image of the quay taken around 1890. In April 1884, the Commissioners received a further letter from Mr Cattle, asking if the railway company could have the use of a strip of land alongside the sidings on the quay, between the gas works to the railway bridge. The Commissioners were agreeable to this and suggested that it should be paved, so as to be suitable for carts loading and unloading the wagons. They indicated that they would pay for this themselves, however the relevant LNWR minute on the subject suggests that it was the railway company who footed the bill.

It will be recalled that, as part of the 1878 agreement between the LNWR and the Commissioners, both parties had the option to extend the existing sidings on St George's Quay beyond the Carlisle railway bridge if they so wished. In June 1886, the LNWR wrote to the Commissioners asking whether they would consider doing this. The latter referred the subject to a sub-committee, who were to confer with Mr Cattle on the matter. There are no further references to this in the Commissioners' minutes and certainly no extension was ever built. There was clearly some interest in the project on both sides. There seems little doubt that the LNWR would have been happy to construct it, but would perhaps have expected the Commissioners to foot the bill. It may be that no agreement could be reached on this point and the whole matter was dropped.

Although there was now a small but steady traffic of iron ore coming into the port, this was almost exclusively from Northern Ireland and carried in small coastal vessels. What the LNWR really wanted was to attract the larger ocean-going vessels carrying Spanish iron ore for Carnforth iron works. It is likely that the LNWR's local goods agent had been trying hard to persuade the ironworks to begin using Glasson Dock for at least some of these larger shipments, and perhaps it was felt that a further reduction in the dues on this commodity might just tip the balance in the railway's favour. With this in mind, in 1884, the railway company made a further approach to the Commissioners asking them to consider another reduction. In November of that year, the Commissioners' clerk visited both Barrow and Fleetwood to ascertain whether there were any special arrangements for the reduction of tonnage dues on vessels engaged in the iron ore trade from foreign ports. It transpired that there were none. Nevertheless, it was decided that the charge for iron ore from foreign ports should be reduced to sixpence per ton on the registered tonnage of the vessel for the first voyage and threepence per ton thereafter, in effect, creating a sort of loyalty bonus. The new rates would be brought in on the 5th November. In spite of this, the LNWR made a request for a further reduction in the rates chargeable on foreign iron ore. However, the Commissioners, having already accommodated the railway company in this matter, politely refused, seeing no reason to make a second reduction.

In fact, the reductions made thus far in the dock dues and what was probably some hard canvassing on the part of the railway company, did, finally seem to be bearing fruit. The Commissioners' shipping register records that in November 1885, the first ship carrying Spanish iron ore arrived at Glasson Dock. This was the SS *Marion Lee* that had sailed from Cartagena. The *Marion Lee* was a screw steamer with a net registered tonnage of 405, still not a large vessel, but significantly bigger than the small coastal vessels employed in the Northern Irish trade. A second cargo arrived in December. However, for the time being, the trade in Spanish iron ore did not develop any further and, in 1886, there was only one cargo, although the small regular shipments from Northern Ireland continued as before.

At this point, with the dock dues on this cargo probably as low as the Commissioners were prepared to take them, the LNWR focussed instead on ensuring that the facilities for unloading the ore at Glasson Dock were operating as smoothly as possible. In March 1885, the Commissioners received a letter from Mr Cattle, requesting the provision of another iron ore bucket to be held in reserve in case of breakage, something that had occurred the previous week. It was agreed that a further bucket be obtained from Alsop Bros of Preston who had supplied the original four buckets. This indicates that the iron ore vessels were unloaded using both cranes, each with two buckets. An empty bucket would be placed in the ship's hold and loaded by hand. Once full, it would be replaced with the second, empty bucket. The full one would, wherever possible, be placed inside the railway wagon and unloaded by a second gang of men. In this way, the operation would be continuous, providing that both the men and the machinery continued to work smoothly. In May 1885, the railway company requested permission to place a hut on the quay at Glasson Dock for the accommodation of an official who was engaged in sampling the foreign iron ore. Permission was granted, subject to a rent of five shillings *per annum.* Then, in December of the same year, Mr Cattle requested permission to

The SS *Colina* at Glasson Dock in 1905. On the 21st February of that year she is known to have arrived at the port from Huelva in Andalusia with a cargo of cork wood for James Williamson & Son. The cargo is being stacked high in what appear to be LNWR wagons, secured with ropes and then sheeted over. In due course the wagons will be taken to the long siding at the rear of Lune Mills and unloaded. Alamy.

erect a larger hut, close to the company's cottages. This was to be for the use of the men unloading the iron ore and other vessels. It was intended that they should take their meals in here and use it for their rest periods during night-time working. This was readily agreed to, again subject to an annual rent of five shillings.

Another cargo that the LNWR wished to attract to Glasson Dock was pig iron. We have already seen that, when the steam cranes were being commissioned, twelve sets of slings and chains were ordered specifically for handling this cargo. By 1886, there were regular and frequent shipments from Harrington, a small port on the Cumberland coast between Whitehaven and Workington. The port was served by both the LNWR and the Cleator and Workington Junction Railway and therefore had ready access to the iron works in the Workington area. At Glasson Dock, the lengths of pig iron would have been loaded, loose, into single-plank wagons. In June 1886, Mr Cattle wrote to the Commissioners, requesting some abatement of the dock dues for vessels trading in pig iron. However, the Commissioners declined, saying that the rates charged for this cargo were already so low that they were unable to make any further reduction. Nevertheless, regular cargoes continued to come in from Harrington until 1895, when the traffic died away. The reasons for this are unclear but after this date there were only very occasional cargoes of pig iron.

In spite of the best efforts of the LNWR, and the slightly reluctant reductions in the dock and harbour dues on the part of the Commissioners, the promise of a healthy and expanding iron ore trade was not initially realised. After 1886, the shipments from Spain ceased. The traffic from Northern Ireland also died away, although there were three small cargoes in 1889. It was not until 1896 that the Spanish traffic returned and from then on, with the exception of 1897, a quiet year, there were regular and frequent cargoes. These were carried in vessels of up to 780 net registered tons and were amongst the largest coming to the port at this time. The Commissioners' records suggest that occasionally there might have been two vessels moored in the dock or at the river quay, one being unloaded and the other awaiting discharge. The steam cranes must have been kept busy and as we have already seen, there is some evidence that at the busiest times, the discharge of iron ore was continued at night. The dock dues for each vessel were calculated by reference to the net registered tonnage and this is the figure that appears in the Commissioners' records. It is not possible, therefore, to calculate the actual weight of individual cargoes. However, at certain periods, the frequency of shipments suggests that the LNWR must have been kept busy, bringing empty wagons to Glasson Dock and taking the loaded ones back up to Lancaster. All of this ore was destined for the ironworks at Carnforth. It is not known what inducements were made to enable the port and the LNWR to capture this traffic.

St George's Quay taken prior to 1891 when a masonry extension was constructed. The array of large buildings on the quayside constitutes the frontage of Williamson's St George's Works. The LNWR's wooden inquiry office is prominent in the centre of the picture. There are a large number of wagons standing in the sidings at the eastern terminus of the quay branch. Amongst them are a number of private-owner coal wagons belonging to Garswood Hall Colliery, which was situated to the south west of Wigan. Courtesy Lancaster City Museums.

There is no evidence that the Commissioners made any further reduction in the dock rate for this traffic, so it would appear that any monetary adjustments (assuming there were any) were made by the LNWR.

On the whole, the relationship between the railway company and the Commissioners was a co-operative and collaborative one, although clearly there were limits on both sides as to the extent of this. The Commissioners focused on maintaining the infrastructure of the port. This included maintaining the buoys and perches in the river, dredging the channels, removing accumulated sand from the wet dock, and keeping the river berths level. The Commissioners also maintained two lighthouses in Morecambe Bay, one on Walney island, near Barrow, and the other at Cockersands, just a short distance from Glasson Dock itself. Once a vessel had docked, any cargo travelling out by rail became the responsibility of the LNWR. However, it seems clear that, at this point, the railway company was doing everything it could to increase the level of traffic at the port, actively canvassing for new customers and encouraging the expansion of existing traffics, providing, of course, that the relevant cargoes would be leaving Glasson Dock by rail. In this, they pulled the Commissioners along with them, requesting additions and improvements to the port's facilities and making regular attempts to persuade them to reduce the dock dues.

At Glasson Dock there were significant areas of open ground, principally used for the storage of various cargoes discharged from vessels. However, the predominant commodity to be seen, at least before the Great War, was undoubtedly timber, mainly in the form of deals, which were sawn, dimensioned lengths of softwood, usually pine or fir. The planks were stacked high on the quayside and this can be clearly seen in the background of many photographs taken around the dock in the nineteenth and early twentieth centuries. The unloading and stacking of the timber was carried out by gangs of men, hired either by the railway or by the recipient of the cargo. Whilst mainly local labour was employed, at busy times additional men were brought over from Fleetwood. There are also references in the Commissioners' minutes to rafts of imported pitch pine logs being towed from Fleetwood and Barrow, to Glasson Dock by sea. One firm in particular, William Huntington and Sons, was a long-established timber importer and merchant, which shipped timber into Glasson Dock for many years. The *Lancaster Gazette* for 6th September 1884 mentions two ships discharging at the dock for William Huntington; the *Normand* from Richibucto, a river port in New Brunswick, Canada, with a cargo of spruce and pine deals; and also the *Ellen* from Fredrikstad, in Norway, with floorboards, ladder poles, etc. Although based in Lancaster, by December 1884, Huntington had also established a timber yard at the dock and, in 1897, he went a step further and established a sawmill on land next to the storage sheds. William Huntington, the firm's principal, was also for many years one of the Lancaster Port Commissioners. This was a common occurrence, with local business users of the port, wishing to exercise some influence over its day-to-day running and development.

With the branch line opened for business, most of the timber imported into Glasson Dock was now leaving by rail and the LNWR was keen to develop this traffic. Having thoroughly probed the Commissioners' defences concerning the dock and harbour dues, the railway felt that perhaps some sort of reduction could be secured in relation to the rates charged for the storage of timber on the quayside. This would make the port more attractive to timber importers, which, in its turn, would increase the amount of traffic leaving Glasson Dock by rail. During the first half of 1884, following some prompting from the railway company, the Commissioners had gone into the question of storage charges on the quays for both deals and for other merchandise. The result was a new schedule of charges and conditions, which was brought into effect on the 1st August of that year. However, the storage rates charged by the LNWR (on the land belonging to the canal) and by the Commissioners differed, the charges levied by the latter being slightly higher. The matter was discussed at a meeting between Messrs Cattle and

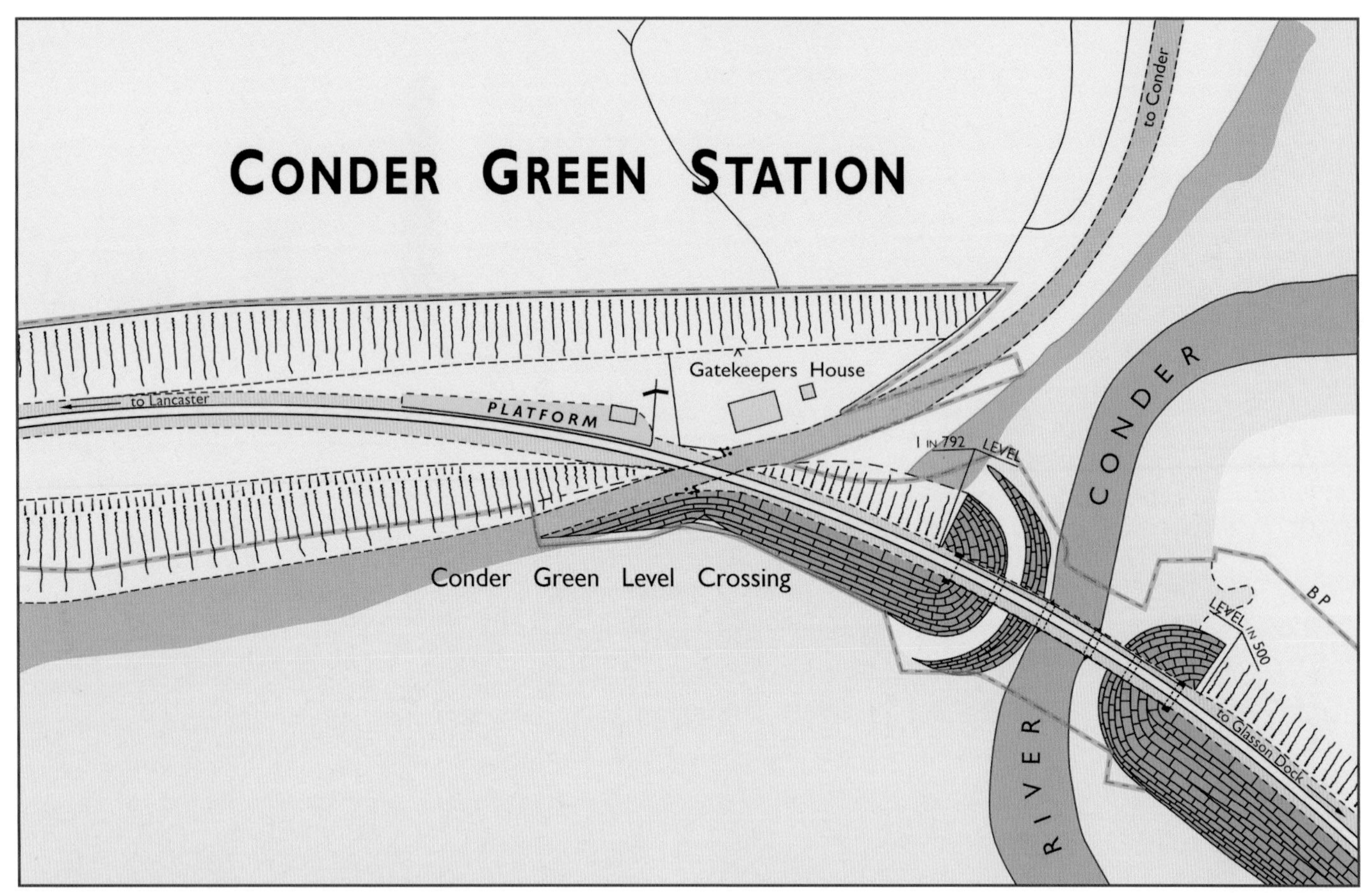

Conder Green Station and the Conder viaduct. As with the plan of Ashton Hall Halt, this is derived from an LNWR legal map dating from the early 1890s. The green lines indicate the extent of the railway's property boundaries.

Map redrawn by Alan Johnstone.

Slinger of the LNWR and the Commissioners, in late November 1884, seemingly without any agreement being reached. On the day following the meeting, the Commissioners received a letter from Mr Slinger with a proposal. He suggested that their storage rate for deals be reduced to half a penny per square yard per month. However, if the deals were leaving Glasson other than by the railway or the canal, for example by road, then there would be an additional charge of 4d per ton of 50 cubic feet. As Mr Slinger explained, the main purpose of these adjustments was to make the storage rates on the Commissioners' ground and on the canal ground uniform. Whilst some reduction in the rates might well attract further timber traffic, it would also be easier to induce other importers to begin using Glasson Dock if a single storage rate could be quoted, rather than having different rates for separate parts of the dock area. The suggestion of a lower rate for rail and canal users suggests that at least some of the stored timber was leaving the port by road, perhaps being carted as far as Lancaster. The LNWR, of course, would have been keen to incentivise the use of the railway or the canal wherever possible. Mr Slinger ended his letter by saying that he trusted that the proposed arrangements would be the means of drawing all parties together and tend to a development of trade by Glasson. However, in spite of Mr Slinger's conciliatory tone, the Commissioners resolved unanimously that their storage rates would remain unchanged.

Turning now to passenger traffic matters, in August 1884, the inhabitants of Ashton, Thurnham and Cockerham, memorialised Mr Cattle, the District Superintendent of the LNWR, requesting the provision of a station at Conder Green. They suggested that the pre-existing crossing keeper's cottage at the eastern end of the viaduct over the River Conder could serve as a ticket office. By rail, this was a mere 45 chains from the station at Glasson Dock, however, owing to the river and the surrounding marshes, a journey by road to the station from the Lancaster side of the river involved a significant detour. The petition coincided with Richard Moon's visit to the branch in August of the same year and his temporary moratorium on any further development of the facilities at Glasson Dock is likely to have put an end to any further action in the matter. Certainly, the subject appears to have dropped from sight until 1887, when the question of a station at Conder Green was taken up by James Williamson, floorcloth manufacturer, owner of Lune Mills and St George's works and also, by this time, Lancaster's Liberal MP. Williamson was also apparently a shareholder in the LNWR but above all, his rapidly expanding business would have been a major customer of the railway in Lancaster. All of these factors seem to have tipped the balance and, in October 1887, the *Lancaster Gazette* reported that work was in progress to construct a small station next to the crossing keeper's cottage at Conder Green. This was opened for traffic on the 5th November 1887. Accommodation consisted of a short, timber platform facing the river, with a wooden hut incorporating a shelter for passengers and a small ticket office with a booking window. However, the latter appears to have fallen into disuse and by the later 1920s, tickets between Conder Green and Lancaster were issued by the guard on the train. The public timetable shows that initially, trains only called at the station on Saturdays and even then, only the first three trains from Glasson Dock called there to pick up and only the last two trains from Lancaster called to set down. The sole purpose of these arrangements was to provide the inhabitants of the rather sparsely populated catchment area, with the opportunity of attending the market at Lancaster on Saturdays.

In September 1888, the Royal Liverpool Manchester and North Lancashire Society's agricultural show was held over a period of three days at Lancaster. The venue was a 22-acre site on Lancaster marsh provided by the Corporation and described in the *Liverpool Mercury* for 6th September as being conveniently situated for the railway stations. At this time, the facilities available for longer distance road transport were somewhat limited. Consequently, the show would have generated a significant amount of additional rail traffic, as animals and equipment were transported from various parts of Lancashire to the county town and then taken away again at the conclusion of the show. In March of that year, the LNWR had discussed the provision of a temporary cattle landing and crane siding, together with a roadway alongside for carts. This was to be situated alongside the Glasson Dock branch. The precise location of this is not known, but it was clearly on the marsh and the rail connection was probably close to the bottom of the gradient from Castle Station. Subsequent development in the area would have eradicated any trace of the site.

Chapter Five

Relative Prosperity

IT IS FAIR TO SAY that, by the end of the 1880s, in spite of the best efforts of all concerned, traffic at the port as a whole had not increased as much as the Commissioners and the railway company would have liked. Comparative year-on-year figures, show that until 1896, when the shipments of Spanish iron ore began coming into the port on a regular basis, there was, overall, very little increase in the registered tonnage using the port. However, this apparent lack of success at Glasson Dock, did not discourage the Commissioners from turning their attention to improving the quay facilities further upriver at Lancaster. The principal reason for this was the rapid growth of James Williamson's floorcloth manufacturing business and, in particular, the considerable expansion of the Lune Mills site. In April 1883, he purchased a large plot of land adjoining the old shipyard site and began to expand Lune Mills into the newly acquired area. Further acquisitions of adjoining parcels of land followed, and one of these included the site of Lancaster's isolation hospital, which was replaced by a new facility on the marsh to the west of Lune Mills. In the longer term, it was James Williamson's intention to develop his business by moving into the manufacture of linoleum, a relatively new product at this time. Linoleum was very similar to floorcloth, the main distinctions being the inclusion of ground cork in the composition or kivver, the universal use of hessian as a backing material and the coating of both sides during the manufacturing process. It was the rapid rise in the popularity of this product in the late nineteenth and early twentieth centuries, together with aggressive marketing and a conscious effort to cater for the cheaper end of the market, that was to make James Williamson a multi-millionaire. It also provided an interesting and distinctive example of Victorian industrial paternalism, played out against the background of a north Lancashire town.

As early as November 1883, Williamson had applied to the Board of Trade to register a decorative design for use on linoleum. However, it was not, seemingly, until 1887 that the product was being produced in quantity at Lune Mills, and the 25-inch Ordnance Survey of 1890 describes the site as engaged in the manufacture of both floorcloth and linoleum. We have already seen that Williamson elected not to have any kind of siding, private or otherwise, running into the Lune Mills site. However, a single siding, the property of the LNWR, extended down the entire length of the site, right alongside the line to Glasson Dock and ended just a short distance from the occupation crossing at Freeman's Wood. It was, in effect, a rather long, extended head shunt for the line running onto St George's Quay. Although there is no firm evidence one way or the other, this must, from the outset, have been used for rail-borne traffic to and from Lune Mills, with access into the works being *via* the back door so to speak. A plan of the works, dating from just after the turn of the century, shows a rear gate on the site, directly opposite the siding, with a weighbridge just inside the entrance. Different editions of the 25-inch Ordnance Survey show what appears to be a loading platform at the rear of the works alongside the siding. Finally, one of Williamson's own plans of Lune Mills incorporating the associated railway lines, and dating from 1902, designates the extended head shunt as a goods and coal siding. However, given that the New Quay was situated literally at the front gate of the site, it was inevitable that at least some raw materials would be brought in by sea. In fact, the Port Commissioners' records indicate that many of the materials for oilcloth manufacture were being brought in by sea and then taken upriver to Lancaster, even before the opening of the railway. These would be discharged either on St George's Quay adjacent to the Carlisle railway bridge, or at the New Quay opposite Lune Mills. However, following the rapid development of the latter site, and the subsequent increase in seaborne traffic, the Commissioners decided that the New Quay required rebuilding and enlarging. Plans and specifications were drawn up and after advertising for tenders, the work was let to Messrs Alexander Morrison & Son of Edinburgh in February 1888. At the same time, the river channel between St George's Quay and the New Quay was dredged to provide deeper water and to level the berths in both places. The final coping stone on the enlarged quay was laid at the beginning of September 1889, although as later as December 1890 the surface was being rolled with a steam roller and plans were being prepared to pave some of the area with stone sets.

Although the enlarged New Quay remained the property of the Port Commissioners, it became, in effect, Williamson's private river wharf. By the early 1900s, the firm was probably the largest single user of the port as a whole. Coasting vessels loaded with china clay, whiting (powdered chalk), resin, white lead, spirits of turpentine and various pigments, negotiated the river channel at the highest tides to discharge their cargoes at Lancaster. None of this travelled by rail of course. However, by looking at what commodities are known to have been consumed at Lune Mills and comparing this with what came in by sea, it is, by a process of elimination, possible to gain a good grasp of the

A rather misty overcast scene at Glasson, which appears to date from the period 1901–1903. Apart from the Coal Tank, the carriages are of considerable interest. The first two are relatively modern 30ft 1in six-wheelers, followed by what appear to be two older 32ft luggage composites. This is not the usual branch set of carriages and it is likely that this is a special or an excursion train.

E Pouteau. L&NWRS.

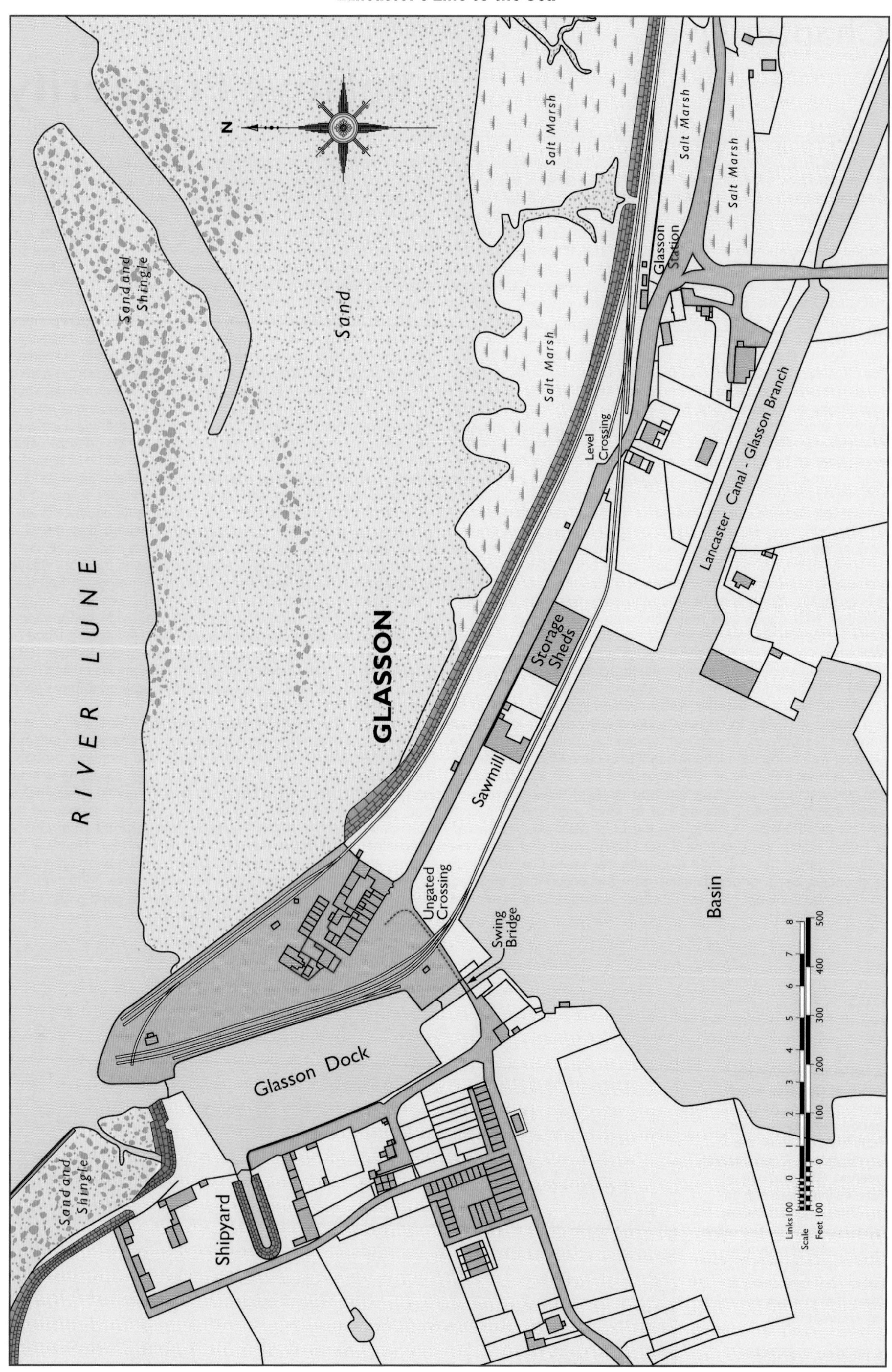
N
RIVER LUNE
Sand and Shingle
Sand
Salt Marsh
Salt Marsh
Salt Marsh
Salt Marsh
Glasson Station
Level Crossing
Lancaster Canal - Glasson Branch
GLASSON
Storage Sheds
Sawmill
Ungated Crossing
Swing Bridge
Basin
Glasson Dock
Sand and Shingle
Shipyard
Scale
Links 100 0 1 2 3 4 5 6 7 8
Feet 100 0 100 200 300 400 500

Another view of the station at Glasson with the stationmaster and just a sprinkling of passengers. The piles of loose material alongside the platform road appear in a number of photographs of the station and may be ash from locomotive fireboxes. It may be that fires were regularly cleaned here whilst the crew waited to make the return trip to Lancaster.
Lens of Sutton Association.

extent to which Williamson's used the railway prior to the Great War. This changed over time and some commodities arrived *via* both modes of transport at different periods. Hessian cloth is a case in point. This was used as a backing for the more expensive grades of floorcloth and later on in the manufacture of linoleum. The cloth was woven in Dundee and in the early 1880s, some of this came in by sea. However, after a short while it largely disappeared from the shipping returns and, therefore, it must have arrived at Lancaster by rail.

In so far as it is possible to discern any sort of a pattern, it would seem that the firm adopted a very flexible approach to the transport of raw materials. As far as Lune Mills was concerned, with the New Quay at the front door and the railway siding at the rear, James Williamson was well placed to respond to changes in the cost of transport and the availability of materials and quickly switch from one mode of transport to the other. Plans of the site show that significant areas were given over to the storage of raw materials, such as cork and linseed oil. These provided a buffer when supply chains were disrupted. This also enabled Williamson to purchase these materials when prices were low and to stockpile them. St George's Works was similarly well served, having access to both St George's Quay and the sidings at the terminus of the quay branch. In the latter part of 1891, Williamson had pressed the LNWR to provide a crane at the quay sidings near to the entrance to the gasworks. This appears to have been put in during February of the following year. In 1903, when a new crane was provided, the opportunity seems to have been taken to increase the capacity to ten tons. All of this suggests that Williamson's and no doubt other port users, were making regular use of the facilities at St George's Quay.

Whilst examining Williamson's transport policies and how they affected the traffic on the branch, there is another point to consider. Irrespective of the nature of the cargo, the size of some of the vessels was such that they were simply too large to come up the river channel to Lancaster. In these instances, the cargo would have to be discharged at Glasson Dock. Then, in the pre-motor age, it would have been loaded into railway wagons for the final part of the journey up to Lancaster. The smaller coasting vessels, of course, could reach Lancaster without too much difficulty, although as with the coal-tip berth at Glasson, there were sometimes problems with maintaining a level riverbed for vessels to rest on at low tide. Clearly, the larger the vessel, the more chance there was of it running aground in the river channel and possibly incurring damage. Depending upon its size, the state of the tide and the amount of fresh water coming down the river, a decision would have to be made, as to whether to attempt the journey upriver or to discharge the cargo at Glasson Dock. When the latter course of action was decided upon, the LNWR might have to respond quickly to ensure that a sufficient number of empty wagons were sent down the branch to be available for loading at the dockside. Speaking in 1900, James Williamson (by now Lord Ashton; he had received a knighthood in 1895) was of the opinion that it was a risk to bring any vessel much above 130 feet long, upriver as far as Lancaster. Nevertheless, it is clear that at this time, vessels of up to 160 feet in length, at certain states of the tide, were making their way to Lancaster to discharge their cargoes for Williamson's at the New Quay.

Glasson Dock, the village and the station as it was around 1910. Nicholson's ship yard is situated at the top left hand corner of the wet dock. The railway storage sheds and Read's sawmill are sandwiched between the canal basin and the road into the village. The passenger station, with its tiny goods yard, can be seen situated further eastwards along the road and quite remote from the village itself. Map redrawn by Alan Johnstone

One commodity, which was consumed in large amounts by Williamson's, but is conspicuous by its almost total absence in the shipping returns until the 1930s, is coal. The only exceptions to this tended to occur briefly during prolonged coal strikes, when some cargoes of Scottish or Welsh coal were discharged at the port. Taken together, the sites at St George's Quay and Lune Mills maintained a large number of steam boilers and, therefore, consumed very significant amounts of coal. As well as utilising steam power to drive the machinery, extensive use was made of steam-heated drying rooms where the newly manufactured linoleum was hung, before being rolled up ready for transport. From the early years of the twentieth century, Lune Mills generated some of its own electricity, but again, the generating plant was steam powered. Although a large body of material relating to Williamson's has been preserved at Lancashire Archives it has

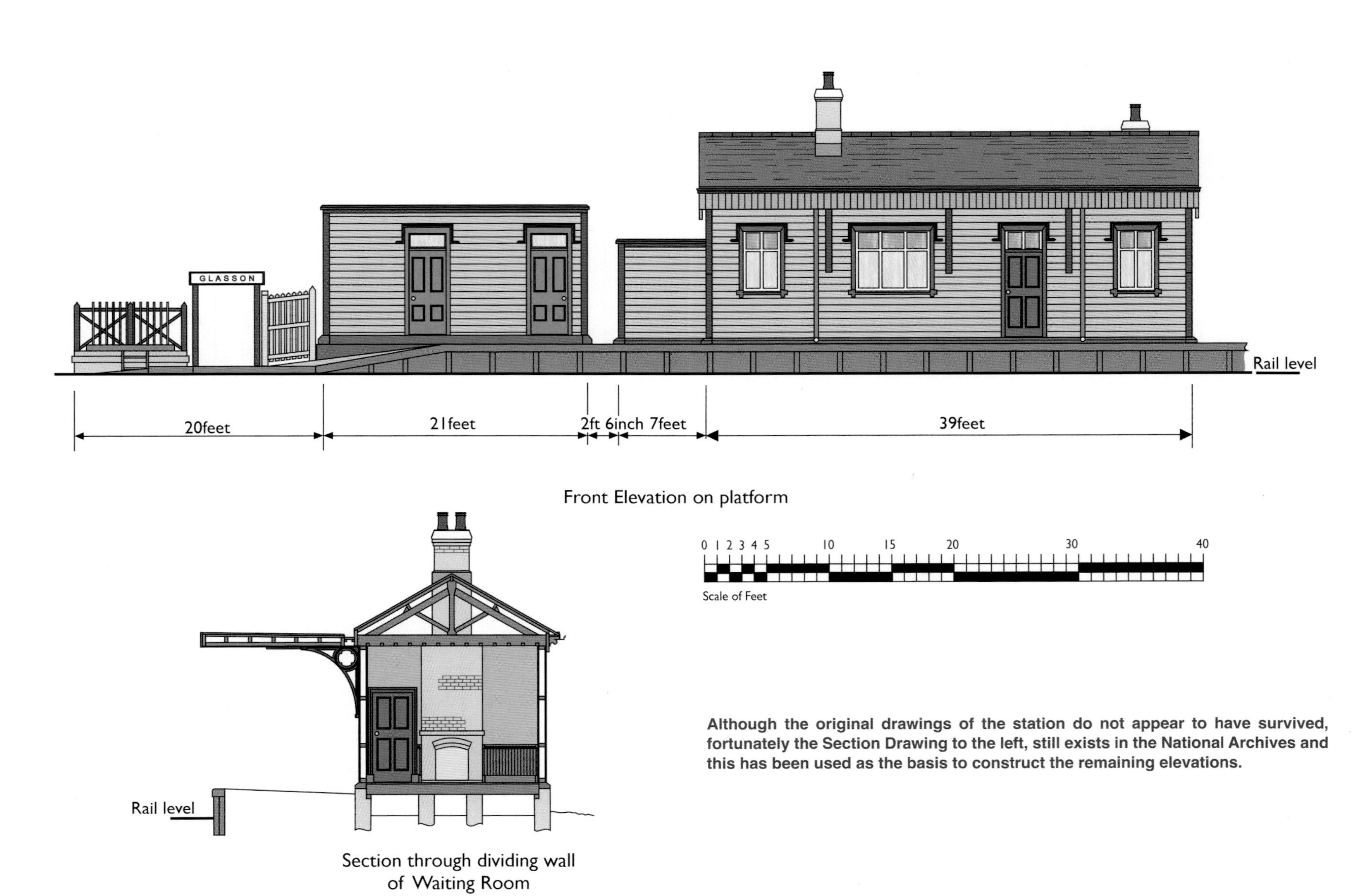
GLASSON DOCK STATION BUILDINGS
GLASSON
Rail level
20feet
21feet
2ft 6inch
7feet
39feet
Front Elevation on platform
0 1 2 3 4 5 10 15 20 30 40
Scale of Feet
Rail level
Section through dividing wall of Waiting Room
Although the original drawings of the station do not appear to have survived, fortunately the Section Drawing to the left, still exists in the National Archives and this has been used as the basis to construct the remaining elevations.
Drawing by Philip Grosse

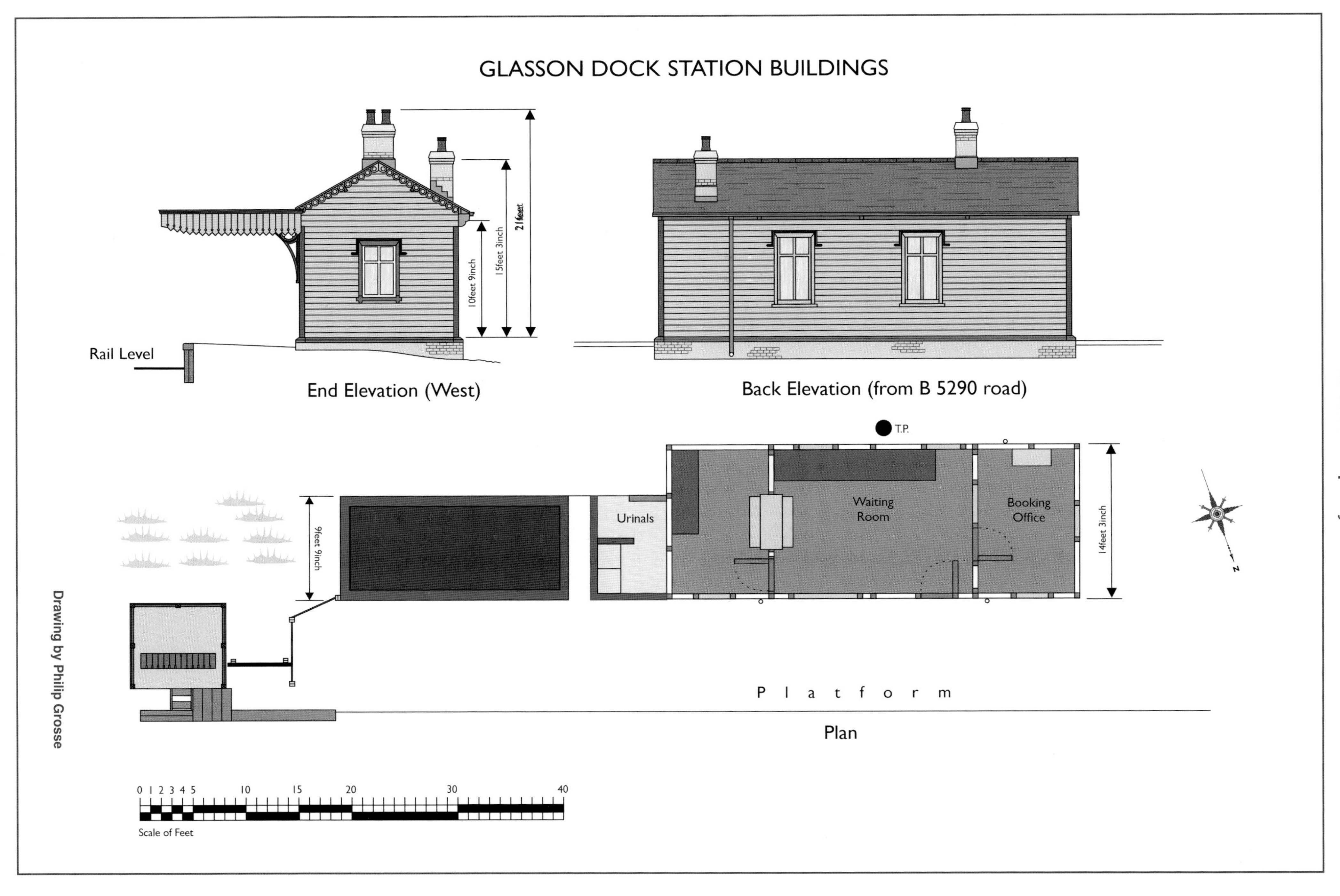

Drawing by Philip Grosse

There is much of interest in this view of Glasson dock taken around the turn of the century. Apart from the dumb buffered wagons from Bickershaw Colliery and one of the Commissioners' steam cranes, on the far left there is an LNWR Wolverton tender dating from the 1860s. It is not attached to a locomotive and it is unclear what its precise purpose is. However, it is likely that it was used to supply coal and possibly water to the steam cranes.
Courtesy Lancaster City Museums.

not proved possible to ascertain from where the firm sourced its coal supplies at this time. A photograph of St George's Quay just downstream of the Carlisle railway bridge, taken at the end of the nineteenth century, reveals a number of wagons from Garswood Hall Colliery which was located to the south west of Wigan. The coal in question may have been for Williamson's, as the wagons are standing right outside St George's Works. However, it could equally have been destined for the gasworks. Certainly, coal for St George's Works was unloaded at the sidings on the quay and then carted into the premises. This was the practice here as late as 1937. Wagons destined for Lune Mills, on the other hand, were placed in the long siding at the rear of the site and the coal was carted into the works from there.

Another cargo received at the port, and particular to Williamson's, was cork wood. This was imported from Spain or Portugal and was ground up in the works and used extensively in the manufacture of linoleum. It was first brought to the port in the early 1890s. However, Williamson's appear to have been using cork prior to this, so up until then, it presumably came in at another port and completed its journey by rail. The vessels themselves were amongst the largest arriving at the port of Lancaster at this time, most of them having lengths of between 200 and 260 feet. Consequently, they all discharged their cargoes at Glasson Dock and it is known that the cork was loaded into railway wagons for the final part of the journey to Lune Mills.

So far, we have only considered the raw materials coming into Williamson's two sites served by the branch. However, together, they produced very large quantities of linoleum, floorcloth and associated products. By 1900, these commanded a large domestic and international market, with strong sales in Europe, Australia and Canada. Although there is no specific confirmatory evidence for the period prior to the Great War, it seems certain that the output from both works travelled to its destination (or to the docks for export) by rail. Finished rolls would be carted from the warehouse and loaded into railway wagons or vans in the sidings at the rear of Lune Mills or on the quay in front of St George's Works. It has been said that the Midland Railway also provided facilities for Williamson's traffic at their Green Ayre station, situated on the other side of the town. Unfortunately, it has not proved possible to discover any evidence to confirm this. However, utilising both railway companies for the carriage of finished goods, or perhaps even for the supply of incoming raw materials, would have enabled James Williamson to play one off against the other in order to achieve the best outcome in terms of cost and level of service. Up in Scotland, the North British Railway provided Williamson's largest British competitor, Nairns of Kirkaldy, with purpose-built, six-wheeled wagons for carrying their rolls of linoleum. However, the LNWR never went this far. Perhaps because, unlike the North British at Kirkaldy, the Euston company was never able to have a monopoly on the traffic.

It should be pointed out that Williamson's was not the only firm in Lancaster manufacturing oilcloth. Storey Brothers & Co Ltd was a smaller concern than Williamson's and their White Cross works was situated away from the river just to the south of the town centre. In fact, the site was fairly close to the terminus of the LNWR's Old Station Yard and, in 1888, the railway company had proposed that a tramway be constructed across what is now the A6 trunk road to connect Storey's establishment with the northern end of the yard and the rest of the railway system. Unfortunately, the Corporation vetoed the idea, perhaps not wishing to have a level crossing on the principal southern route in and out of the town. Nevertheless, White Cross works was situated directly alongside the Lancaster Canal and some commodities were brought to the site using this method of transport. Certainly, in the years immediately after the end of the Second World War, barges were still bringing coal up the canal and unloading it directly into the works. The Commissioners' shipping records reveal that, for many years, sporadic cargoes of whiting and china clay for Storey's were received at the port.

However, given the size of the vessels involved, it is likely that they were brought up to Lancaster to be unloaded there and then carted to the White Cross works.

We turn now to consider the rail traffic generated by the Corporation gas works situated on St George's Quay. For some years, the gas works appears to have received its supplies of coal by sea. However, this arrangement ended once the quay branch was opened for traffic. In 1883, the year the line opened, the site consumed 5,952 tons of coal. However, by 1890, the annual consumption had increased to approximately 12,000 tons of coal and cannel, the latter being a hard variety of coal particularly suitable for the manufacture of gas. At this time, a colliery coal wagon had a capacity of between eight and ten tons, thus providing a very rough estimate of 1,400 laden coal wagons coming into the gas works in that year, with a corresponding number of empties returning up to the main line at Castle Station. This significant increase in the consumption of coal suggests something like a doubling of gas production and consumption within the town in the intervening seven years. Following their takeover of the gas works in 1880, the Corporation began to upgrade and expand the gas producing plant and also made a significant effort to advertise and to popularise the use of gas within the town. There seems little doubt that the provision of the private railway siding into the works, facilitated this policy of expansion.

It's a rather cold day in February 1895 and this image shows to good effect the original timber face of the Ford Quay together with the new masonry extension. There are some low-sided and bolster wagons on the sidings, which appear to be carrying baulks of timber. Lancaster Corporation's gas works is on the left and on the right the steaming, smoking bulk of Williamson's Lune Mills rises out of the mist.

Courtesy Lancaster City Museums.

It is known that, in 1900, the Corporation accepted a tender for the supply of 5,000 tons of screened Wigan 4ft gas coal from Messrs Cross Tetley & Co. This concern worked Bamfurlong and Mains collieries situated to the south of Wigan. The coal was to be delivered over the twelve months commencing 1st September 1900. In 1914, the Corporation Gas Committee had three coal contracts with different suppliers; firstly, with Crawshaw & Warburton's Shaw Cross colliery, supplying 10,000 tons and situated on the Great Northern Railway's Ossett branch; secondly, with James Mitchell, of Todmorden, who supplied 5,000 tons; and finally, there was a contract for 7,500 tons with Samuel Thompson & Co of Lancaster, who we have already come across in connection with coal shipments from Glasson Dock. It seems that by 1922, if not earlier, the gasworks was operating its own railway wagons and that an indeterminate number were acquired from the Ince Waggon & Ironworks Co in that year. In 1938, one

Class 2MT No 46422 is seen here shunting on the Ford Quay. The train of fitted vans is slowly being pushed across New Quay Road and is probably destined for the siding running into Williamson's dispatch warehouse. The photograph was taken in March 1963.

Ron Herbert.

This wagon was constructed in 1922 by the Ince Waggon & Ironworks Co and is seen here following a repaint in 1938. The new livery appears to be black with white letters shaded red. The instructions on the side of the wagon relate to Crawshaw & Warburton's Shaw Cross Colliery, however, it is known that in 1938 the gas works also obtained coal from James Mitchell of Todmorden and local coal factors, S Thompson & Co.

HMRS.

or more of these was repainted in a new and rather striking livery. In that same year, the gasworks carbonised 25,159 tons of coal, representing a significant increase in the consumption of gas in the town since 1890.

Another regular user of the port was James Cropper & Co, which operated a group of large paper mills at Burneside (between Kendal and Windermere). The company made extensive use of the railway for the transport of raw materials and the dispatch of finished products. Indeed, Cropper's had their own private tramway connecting the mills with the goods yard at Burneside station. The company began bringing significant cargoes of wood pulp into Glasson Dock around 1891 and these continued until the Great War. Cropper's had been using this material for the manufacture of paper since the 1880s and, by 1895, it had superseded jute and Surat cotton as the principal raw material for their products. In the earlier years, the size of the vessels was such that it is possible that, at certain states of the tide, some of them made their way up to St George's Quay, to discharge their cargoes into railway wagons at the sidings there. However, given the superior unloading facilities in the shape of the two steam cranes, Glasson Dock might still have been the preferred option. Certainly, by the early years of the twentieth century, the size of the pulp boats had increased to the extent that they would have to have been discharged at Glasson Dock. There were also, from time to time, cargoes of china clay destined for Cropper's. These were carried in smaller coasting vessels and consequently, it is possible that the cargoes were taken upriver and transferred into railway wagons at Lancaster. However, as with the shipments of wood pulp, discharge at Glasson Dock is perhaps more likely.

After successfully completing the reconstruction of the New Quay, the Commissioners immediately turned their attention to the portion of St George's Quay immediately downstream from the Carlisle railway bridge. This, of course, was served by the eastern end of the Lancaster Quay goods line. By 1890, the existing facility was deemed to be inadequate for the level of traffic now presenting itself, therefore in December of that year, it was decided to extend the existing quay wall westwards by 500 feet. The new work was to be executed in masonry. Once again, plans and specifications were drawn up and, in April 1891, the contract was let to Mr William Harrison of Moor Lane, Lancaster. There were already at least two structures on the quay, the inquiry

An unidentified vessel is being loaded with bunker coal by the Port Commissioners' two cranes. The coal appears to originate from New Ingleton Colliery situated just over the border in Yorkshire and served by the Midland Railway. In the period 1914 to 1936, this was the nearest significant colliery to Glasson Dock. In addition to the colliery's own wagons, it appears that a number of the Midland's wagons have been pressed into service in order to expedite delivery to the quay side.

Courtesy Lancaster City Museums.

View looking towards Lancaster from the end of the Conder viaduct. The building on the right is the crossing keeper's house which, for a little while, doubled as the ticket office for Conder Green Station. Beyond the gates and associated fencing, the long shallow mound adjacent to the line denotes the site of the station platform.

CRA, Rev J Jackson Collection.

office belonging to the LNWR and a substantial wooden shed, the property of the Commissioners. At this point, the shed was being let to the Whitehaven firm of Wilson and Kitchen for the storage of cement. The final coping stone on the extension was lowered into place on Tuesday 27th June 1893. The older timber-faced quay and its masonry extension now constituted a continuous run of quayside, west of the Carlisle railway bridge. To avoid confusion between the new combined facility and the New Quay opposite Lune Mills, the former was renamed Ford Quay and is referred to in this manner in subsequent Commissioners' minutes. In 1902, it was decided to provide a new shed for the storage of cargoes on the extended quay. In order to save money, the Commissioners decided to dismantle the existing shed and use the timbers to create a new building. The Commissioners' minutes suggest that by 1904, the space within was let to Henry Tyrer & Co, Liverpool shipowners, and Messrs W & J Pye, a local firm of millers and animal feed suppliers. Pye's were very regular users of the port, being based in Lancaster but also having a mill at Thurnham, just a short way up the Lancaster canal from Glasson Dock. Seaborne cargoes, therefore, were discharged at both places.

As part of the continuous expansion of the Lune Mills site, by 1890 James Williamson had established a narrow-gauge railway system within the works for moving raw materials and finished products around the different departments. It is not known whether

James Williamson & Son's internal railway system on the New Quay at Lancaster. Locomotive No 2 pauses before hauling a train of hopper wagons into the works. The white material in the wagons is either whiting (powdered chalk) or china clay that has been discharged from the vessel in the background.

Courtesy Lancashire Archives.

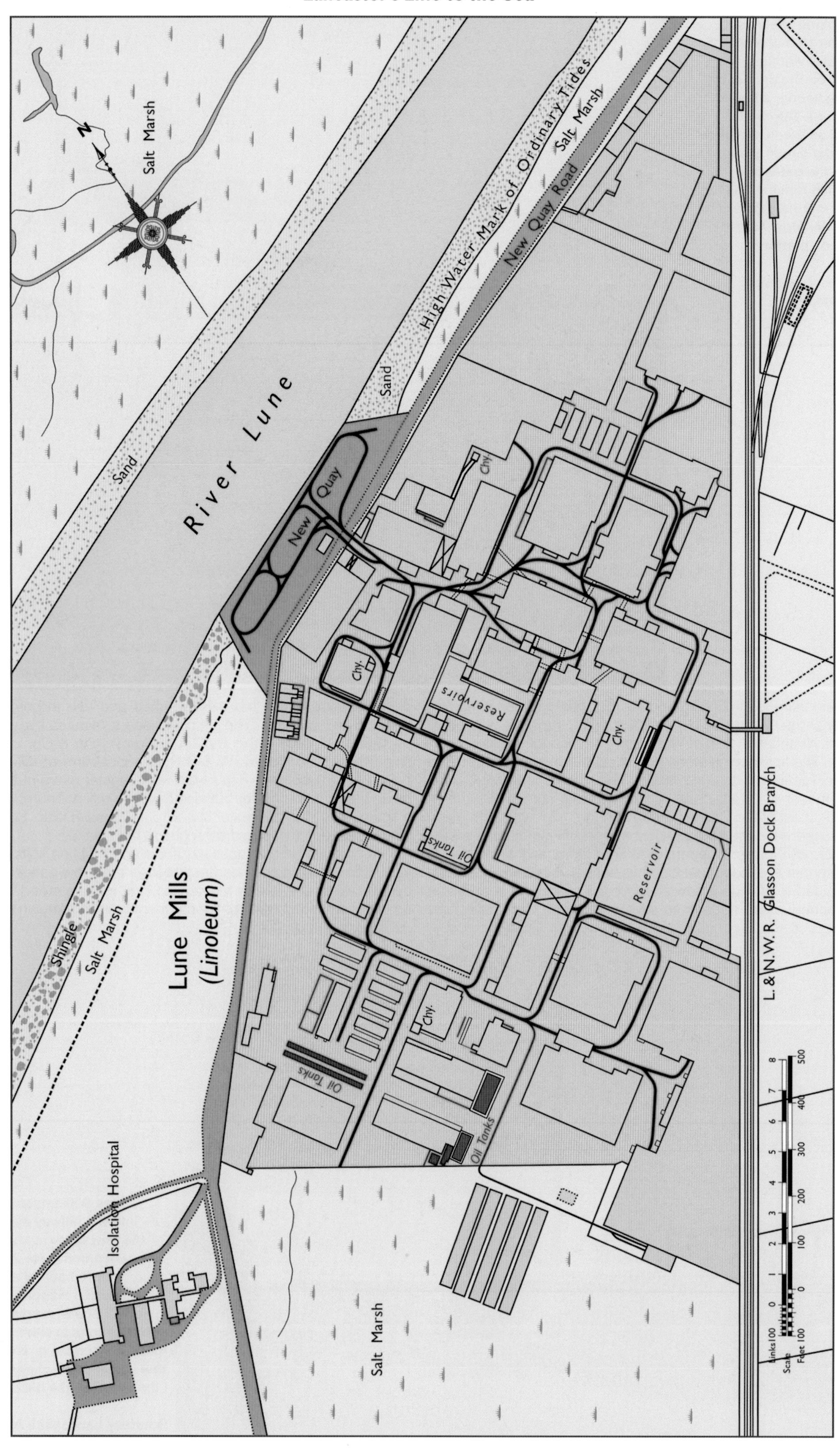
Salt Marsh
N
Sand
River Lune
Sand
High Water Mark of Ordinary Tides
Salt Marsh
New Quay Road
New Quay
Chy.
Chy.
Reservoirs
Chy.
Oil Tanks
Reservoir
Chy.
Oil Tanks
Oil Tanks
Lune Mills
(Linoleum)
Shingle
Salt Marsh
Isolation Hospital
Salt Marsh
L. & N.W.R. Glasson Dock Branch
Links 100
Scale
Feet 100
0 1 2 3 4 5 6 7 8
0 100 200 300 400 500

Discharging either whiting or china clay at the New Quay in Lancaster in the late 1920s. The vessel is one of Robert Gardner's fleet, the *Mountcharles*. Gardner was based at the port and his vessels were a common sight in the river and at Glasson Dock for many years from the early 1920s onwards.

Courtesy Lancashire Archives.

locomotives were used at this stage, but in the absence of any confirmatory evidence it seems unlikely. Trains may have been pulled by horses, with individual wagons perhaps being moved by hand. In May 1900, Lord Ashton wrote to the Port Commissioners, requesting permission to lay what was described as a bogie tram line across the New Quay and into Lune Mills for conveying raw materials being discharged from ships. After some negotiation, the rental for this was agreed at £5 per annum and the final agreement and plan of the arrangements was signed by the Commissioners in January 1904. The final layout of the tracks on the quay was quite ingenious, comprising three distinct circuits, with separate routes onto and off the quay for empty and full wagons respectively. This provided for maximum flexibility and efficient discharge of cargo from the small coastal vessels generally employed for Williamson's traffic. A comparison of the 1890 Ordnance Survey with that for 1910 reveals that, as the Lune Mills site expanded, the railway system was extended and altered to accommodate this. The Williamson's deposit at Lancashire Archives includes several plans of the site from different dates. Unfortunately, none of these provide any details of the internal railway system. At some point, possibly around 1910, Williamson purchased two locomotives from Dick, Kerr & Co of Kilmarnock. Unfortunately, the business records of the company are incomplete and no details relating to these locomotives have survived. However, there is a photograph taken on the New Quay in the 1920s, which provides a clear view and reveals that they were 0-4-0 saddle tanks of a modern appearance. They were unnamed but numbered, predictably perhaps, 1 and 2. By this time, the gauge of these lines was 2ft 9in and this may well have been the case from the start. A 1930 inventory of the assets at Lune Mills, reveals that at this point the company possessed two steam locomotives and two petrol driven ones, with two separate locomotive sheds and one repair shop. The photograph taken on the New Quay during the 1920s, also shows steel tipper wagons being used to transport either whiting or china clay from the quayside into the works. However, the only items of rolling stock listed in 1930 are a number of "steel tray wagons" each 9ft long, 5ft wide and 12ins deep. Photographic evidence indicates that, in 1935, the internal railway system was still in use within the works, seemingly on a daily basis. However, local memory recalls that the locomotives were no longer used after the early 1930s and remained in their shed until sold for scrap around 1935. The lines on the New Quay seem to have fallen into disuse around this time with the company switching to motor lorries to move raw materials from the quayside into the works. It has been said that a branch of the narrow gauge system ran underneath the Glasson Dock branch and was used to tip waste and ash from the boilers onto empty land on the opposite side from Lune Mills. However, it has not proved possible to discover any evidence to support this. That is not to say that such material was not carted across the line and dumped in this area. Indeed, there is some photographic evidence from the 1940s to suggest that this is what happened. Various editions of the 25-Inch Ordnance survey also indicate that, for many years, there was an internal narrow gauge railway within St George's Works on the Ford Quay. It has not been possible to uncover any further information relating to this but, given the extent of the trackwork, it seems unlikely that locomotives were ever used.

All the improvements associated with the quays at Lancaster were a response to the increased number of vessels coming upriver and the enhanced facilities must also in their turn have encouraged further development of the river traffic. At the same time, as we have already seen, the active promotion by the LNWR of the traffic in timber, pig iron, wood pulp, cork and above all the Carnforth iron ore did much to increase the level of activity at Glasson Dock. Although the largest sources of traffic on the branch have been considered in some detail, it is difficult to ascertain the extent to which vessels discharged infrequent or one-off cargoes into railway wagons, especially if they came upriver to the Ford Quay. Shipments of granite sets, cement, macadam, iron pipes, broken stone, oats, corn and maize (for consignees other than W & J Pye) and general goods, all arrived at both Glasson Dock and Lancaster from time to time or sometimes more regularly. Certainly, the limited number of photographs of the Ford Quay taken in the late nineteenth and early twentieth centuries, show significant numbers of railway wagons in the sidings there. However, this may have been largely down to the activities of the gas works and Williamson's St George's works. It is likely that much of what was discharged from vessels at the Ford Quay was simply carted away to consumers within the town and the wider area.

Although incoming laden vessels kept Glasson Dock relatively busy, once the outbound coal shipments had dwindled away, there were few outgoing cargoes available. Up the river at Lancaster, the gasworks on Ford Quay occasionally loaded small vessels with coke or pitch, the former going to Northern Ireland and the latter to Antwerp in Belgium. However, most seem to have left Lancaster or Glasson Dock empty. This inability to procure return cargoes prompted some criticism of the port and may have provided a disincentive for some masters and owners.

Lune Mills and the New Quay circa 1910. The lines in black represent Williamson's internal narrow gauge railway system and show the arrangement of the lines on the quayside. The Glasson Dock branch and the parallel quay goods line run along the bottom of the map. Part of the sidings serving the District Engineer's yard can be seen bottom right. On the far right, the line running onto the Ford Quay can be seen branching off the goods line. Map Redrawn by Alan Johnstone

Taken in the years before the Great War, this view shows an array of vessels in the wet dock. The two tracks on the left date from the opening of the line, whilst the line on the right was installed in 1903 in response to increasing rail traffic at the port. The Commissioners' two steam cranes can be seen in the distance. Courtesy Lancaster City Museums.

The increased activity at Glasson Dock prompted the Commissioners to consider the provision of a third line of rails alongside the wet dock as well as a third steam crane. Towards the end of 1899, there had been some discussion about this between the Commissioners and Lewis D Price, the LNWR's new District Traffic Superintendent. He had been appointed in June of that year following Mr Cattle's retirement. Like his predecessor, he was based in Lancaster and became the Commissioners' first point of contact for all railway matters. The provision of a third set of rails would facilitate the discharge of two ships simultaneously, with the steam cranes utilising the line nearest to the quayside and separate rakes of wagons being accommodated on the inner lines of rails. During the course of a meeting, Price had assured the Commissioners that both matters were receiving consideration and being pushed forward. However, no further action seems to have been taken in the matter on either side until September 1902, when a letter was sent to the LNWR, again pointing out the necessity for a third line of rails, and requesting that the railway company carry out the work on the Commissioners' behalf. There was a further meeting on the subject, with Mr Price confirming that plans for the third set of rails had been agreed and returned to Euston and he was merely waiting for authority to commence the work. However, it seems that construction did not actually get underway until October 1903, some twelve months later. Photographic evidence indicates that, by the early years of the twentieth century, a third crane had been provided at Glasson Dock. However, unlike the other two cranes, this was never the property of the Commissioners and, therefore, must have been provided by the railway company. There is a photograph, taken probably in 1904, showing three cranes working on the quay. Two of them appear to be the Commissioners' dating from 1883, but the third is to a different, slightly larger design. This can only have been supplied by the LNWR.

So as 1903 drew to a close, the traffic at the port of Lancaster looked reasonably healthy. There were plenty of smaller vessels making the journey up to Lancaster, especially to the New Quay with traffic for Williamson's. At Glasson Dock itself, in addition to the coastal trade, there were regular cargoes of Spanish iron ore, together with a steady trade in timber from Canada and the Baltic and periodic shipments of wood pulp from Scandinavia. These were supplemented with occasional cargoes of cork wood from Portugal or Spain. These four commodities, together with the products from Williamson's two sites, formed the staple outbound goods traffic on the line. However, this happy state of affairs was destined to change in 1904 when the Midland Railway opened their new, modern and well-equipped port at Heysham, almost literally just around the corner from Glasson Dock.

Chapter Six

A Fly in the Ointment

A busy scene at Glasson Dock probably in 1904. The Commissioners' two steam cranes are busily engaged in unloading the far vessel. The nearest crane appears to be an extra one brought in by the LNWR to deal with the shipping traffic. The more distant vessel has been tentatively identified as the SS *Lillian* which is known to have come to Glasson Dock in January 1904 with a cargo of iron ore from Spain.
Courtesy of Lancaster City Museums.

THE DERBY-BASED Midland Railway had operated the harbour facilities at Morecambe for many years. As we have already seen, these were constructed by the North Western Railway following the opening of their line from Lancaster in 1848. As from 1st January 1859, the North Western had been leased to the Midland Railway, the terms of the lease being such that the line was virtually vested in the Midland. By the early 1890s, the port facilities at Morecambe were proving inadequate for the level of traffic being presented, furthermore there were ominous signs that the harbour was beginning to silt up. It had originally been planned to make a fresh start and construct a new jetty at Heysham just three and a half miles south of Morecambe. However, by 1895, the Midland Railway had decided to go much further and construct a new, large, fully equipped port. This would be served by its own railway branch line with a dedicated passenger station and extensive sidings to handle the traffic that was expected to pass through the new facility. There would also be modern appliances for cargo handling together with ample accommodation for storage. The relevant Act of Parliament was passed in 1896 and the port began to take traffic in 1904. The new harbour opened out into what is known as Heysham Lake, a broad, deep-water channel leading out into the Irish Sea. Consequently, ships could enter or leave the port at all states of the tide.

The opening of the port at Heysham in 1904 did not presage a sudden defection from Glasson Dock. Indeed, the Commissioners' records show that for 1904, overall, there was a slight increase in the amount of registered tonnage using the port of Lancaster as a whole. However, the figures for 1905 reveal a different story, with a reduction of around a quarter in this figure. The main casualty was the iron ore traffic and, whilst this did not immediately cease, it experienced a very significant reduction. The Port Commissioners countered by applying a one-third rebate on port dues for ships entering Glasson Dock from foreign ports. This was effective from 15th May 1905 and applied to the vessels carrying iron ore, of course, but also to those carrying timber, wood pulp and also to Williamson's consignments of cork wood. However, by 1906, the iron ore shipments were down to only three and by 1907 they were down to two and in 1908 there was only one cargo of this commodity, a stark contrast to the busiest period just before the turn of the century when, in some years, the number of iron ore shipments reached double figures. Williamson's cork boats ceased to use Glasson Dock after 1906, although the firm continued to make almost continuous use of the facilities at the New Quay and, to a lesser extent, at the Ford Quay. The cork was now discharged at Heysham and then presumably brought by rail, in the Midland Railway's own wagons, down the Glasson Dock branch to Lune Mills. If this was the case, then at least the LNWR did not lose the benefit of the traffic altogether. The Port Commissioners do appear to have been very much alive to the situation, and seem to have done all they could to stem the defections to Heysham. As early as August 1904, there had been some communication between them and Lord Ashton about the continued use of Glasson Dock by the cork boats, the latter promising to do all that he could for the port. In October 1905, the Commissioners were seeking interviews with Ashton about the cork traffic and also with a Mr Barton of Carnforth ironworks about the iron ore shipments. The outcome of these interviews is not recorded, but the disappearance of these sources of traffic from the port shortly afterwards indicates that they were unsuccessful. During 1905, there is also evidence of dissatisfaction on the part of both the Commissioners and ship owners with the way the LNWR was handling the traffic to and from Glasson Dock. In May of that year, a deputation

A view of Glasson station in the first decade of the twentieth century. The unidentified LNWR Coal Tank appears to be hauling the empty branch set of six-wheel carriages and may be venturing onto the line to the river quay to retrieve wagons for the return journey to Lancaster. Meanwhile, the passengers wait patiently for their train to be shunted into the platform. In the background, the railway's storage sheds are clearly visible.

Ken Nuttall Collection.

from the Commissioners met with Mr Price of the LNWR to discuss the various complaints relating to the discharge of vessels. Price assured the deputation that all the complaints would be investigated and, if possible, remedied, and that the LNWR would do everything possible to cultivate the shipping trade at Glasson Dock. During the Spring of 1906, there were further discussions on the subject between the Commissioners and the Mr Price with the latter once again promising to do all that he could to being trade to Glasson Dock. Mindful of the limitations of the accommodation at the wet dock in early 1908, the Commissioners approached Messrs D & C Stevenson, engineers of Edinburgh, enquiring about the feasibility and cost of widening the sea entrance to 50 feet. Having received a reply from Stevensons, the Commissioners approached the LNWR to ascertain whether they would be prepared to make a financial contribution towards the proposed improvement. However, the LNWR's traffic manager quickly wrote back saying that his company regretted that it could not see its way to doing this. Nevertheless, the Commissioners' minute books indicate that some sort of dialogue relating to improvements at Glasson Dock continued throughout the rest of Spring that year. Unfortunately, nothing concrete appears to have come of these discussions.

The difficulties of using Glasson Dock and the port of Lancaster are illustrated in an incident that took place in December 1906. The vessel in question, the SS *Queen's Channel* was not a large one. She was registered at only 128 tons and was 158 feet long. She had been due to sail up to one of the quays at Lancaster on the 20th but came instead to Glasson Dock, the master fearing that he would be stuck at Lancaster for several days owing to it being a neap tide on the 22nd. The LNWR was unable to supply the necessary wagons at Glasson Dock to enable the cargo to be discharged. The vessel was consequently ordered by the owners to Heysham, where she quickly discharge her cargo and was able to sail from the port on the following day. In the final analysis, Glasson Dock was a small eighteenth century port with a couple of steam cranes, a coal tip and a railway line. It could not hope to compete with the modern, extensive facilities and round-the-clock access available at Heysham. However, there must also be some criticism of the LNWR for not being sufficiently agile to arrange for the wagons to be available at Glasson Dock, when they were required. This resulted in a loss of business for both the port and the railway and the incident could only further emphasise that the facilities at Glasson Dock were in every way inferior to those at Heysham. There is also perhaps a sense in all of this that the LNWR itself realised that the little port at the mouth of the river Lune would never be able to compete with its larger, more modern, near-neighbour and in effect moved from a policy of promotion to one of managed decline.

As regards the Spanish iron ore traffic, the LNWR appears to have adapted quickly to the loss of this business to Heysham. The Lancaster and Carlisle District working timetable, effective from February 1911, shows that up to six daily, conditional trains of loaded mineral wagons were scheduled between Heysham and Carnforth, with up to six corresponding trains of empties in the opposite direction. These were run in response to vessels discharging iron ore at Heysham destined for Carnforth Ironworks. All of these trains were worked between Heysham and Carnforth, in both directions, by Midland locomotives and men. From Heysham, they reached LNWR metals by reversing direction at the Midland's Morecambe station. The LNWR, of course, would have been able to charge the Midland something for allowing the latter's trains to run on their metals between Morecambe and the ironworks at Carnforth. It was merely a pragmatic acceptance of the *realpolitik* of the relationship between the two competing railway companies in the Lancaster area.

In spite of the loss of both the iron ore and the cork traffic, all was not gloomy at the port. After a significant drop in the registered tonnage in 1905, overall, the level of traffic remained stable with the annual net tonnage figure hovering around the 30,000 mark. It was to remain like this until the Great War. The one-third rebate on foreign cargoes being discharged at Glasson seems to have stimulated the timber traffic, at least for a little while, with some new names appearing amongst the consignees listed in the register of dock dues. Amongst these, in 1907, was the Lancaster Railway Carriage and Wagons Works. In 1908, seven substantial cargoes of timber arrived at Glasson Dock, most of which is likely to have left the port by rail. In the same year, Cropper's discharged three large cargoes of wood pulp at Glasson Dock, which would certainly have travelled

This view, taken around 1948, shows Fowler 4F 0-6-0 No 4032 moving slowly alongside the canal basin bringing two wagons from the Commissioners' sidings towards the station. On the right is the sawmill which, from 1910, was operated by the Westmorland Cooperage and Sawmills Co. For many years this provided a significant amount of traffic for the Glasson Dock end of the branch. However, by 1948 it is likely that the firm had moved over to road transport.

CRA, Rev J Jackson Collection.

by rail to Burneside. Further upriver at Lancaster things were busier than ever. Notwithstanding the defection of the cork boats to Heysham, James Williamson does appear to have kept his word. The number of vessels now visiting Lancaster with cargoes for Williamson's would suggest that, with the exception of coal, the firm was now using shipping wherever possible to bring raw materials into the St George's and Lune Mills sites. However, this would most certainly not have been an act of altruism. For bulky cargoes, such as china clay and whiting, shipping was often a cheaper option than transport by rail. Furthermore, the two works at Lancaster were ideally placed to receive raw materials directly into storage, straight from the ship's hold. Other businesses were also taking advantage of the extended facilities at the Ford Quay. In 1907, Messrs Blythe and Sons, a firm of Liverpool corn millers, erected a shed for their own use on the quay and with this completed, began sending ships to Lancaster on a regular basis. Almost immediately they requested additional space in order to extend their premises and this was granted by the Commissioners. At the same time, Blythe's made enquiries about acquiring their own dedicated siding space on the quay. In 1908, the Commissioners themselves constructed an additional shed on the Ford Quay for the use of the Lancaster and Liverpool Shipping Co. The firm was charged an annual rent of 8% of the cost of the warehouse together with an annual ground rent of £20. At the same time, the LNWR requested permission to extend their inquiry office and this appears to have been granted. Also, around this time, the railway company began using a travelling crane on the quay, although it is unclear whether this was steam or hand powered. In December 1907, they approached the Commissioners with a view to obtaining retrospective permission for its use on port property. All of this indicates that traffic on the Ford Quay was growing significantly during this period. Furthermore, the involvement of the LNWR in this expansion strongly suggests that this was reflected in an increase in railway traffic both to and from the quay. The Ford Quay must have been a busy place in the years up to the Great War. In addition to the shipping and railway traffic, by 1912, Williamson's are known to have been using motor lorries with trailers to move discharged cargoes to their works. Indeed, a driver's error resulted in one of the trailers going over the quay side and into the river. In response to this heavier motor traffic, the western end of the Ford Quay was paved with sets. On the eve of the Great War, there were three sheds on the Ford Quay. The oldest was now solely occupied by W & J Pye, the Lancaster-based corn millers, under a lease dating from 22nd

This charming image shows four generations of the Robinson family on the station platform at Glasson. Michael Robinson was the stationmaster from 1901 to 1909. Prior to this he had been the LNWR's canal agent at Glasson. Michael's son Ernest, who was a railway porter, is standing on the left, and Michael's father, Robert, is seated holding Michael's grandson, Cyril.

Courtesy Lancaster City Museums.

A view of the old watch house on the pier head at Glasson Dock. There are two items of LNWR manufacture on display here. The first and most obvious is the 8-ton van with its sliding roof door. The chalked inscription on the side indicates that there is a problem with the brake and probably explains why it has been shunted out of the way onto the pier head. The other LNWR item is the slotted signal post purchased by the Port Commissioners and used for sending messages to the river pilots at Sunderland Point across the river.

Courtesy Lancaster City Museums.

December 1909. Messrs Blythe & Son occupied their own shed and Messrs Sellars & Son had a lease on the Commissioners' second shed. The latter firm had taken over the Liverpool–Lancaster trade previously carried on by the Lancaster and Liverpool Steamship Co. Incidentally, in November 1909, Pye's had also entered into an agreement with the LNWR to take over the tenancy of a portion of one of the storage sheds at Glasson Dock, thereby cementing the firm's commitment to the port of Lancaster as a whole.

A further sign of an increase in traffic on the quay goods line was the provision, in 1910, of a crossover connecting the Glasson Dock route with the quay goods line. This was situated at the bottom of the gradient up to Castle station, just to the west of where the line onto the Ford Quay diverged from the line down to the Freeman's Wood end of the sidings. The rather minimalist layout of the sidings serving the Ford Quay and Lune Mills and, in particular, the absence of a loop to enable engines to run around their trains, must always have created some difficulties for those tasked with the shunting of the quay goods line. These problems are likely to have become more acute as traffic increased after the turn of the century and, therefore, the crossover must have helped to mitigate the situation. The new facility enabled a train of wagons or a light engine to come from Castle Station on the Glasson Dock route and then reverse, to gain access onto the quay goods line. This was especially useful if the gradient on the upper part of the goods line was already occupied by standing wagons left there during the course of shunting operations. It may have been that, from time to time, the engine working the goods line simply became blocked in by the sheer numbers of wagons and it proved necessary to send a second locomotive down *via* the Glasson Dock route in order to assist. The new crossover also enabled goods trains from Glasson Dock to run directly onto the quay goods line without having to run up into Castle Station in order to reverse direction. This might occur from time to time when, owing to the size of the vessel or the state of the tide, cargoes destined for Lune Mills or St George's Works were discharged at Glasson Dock instead of being brought upriver. The crossover was controlled by a two-lever ground frame, which was unlocked by the staffs for both the Glasson Dock line and the Lancaster Quay goods line. The new work duly received a visit from the Board of Trade's inspector, Colonel Yorke, on the 16th September 1910, who pronounced everything to be satisfactory.

Glasson canal basin. The fisherman's repose is disturbed as Fairburn tank No 42063 moves slowly along the grass-grown single line with a train of open wagons. This was taken on 10th September 1963 — the locomotive was withdrawn the following month. The stone wall on the far right is part of what was once the sawmill, operated from 1910 to 1954, by the Westmorland Cooperage and Sawmills Co.

Derrick Codling.

Ivatt 2-6-0 No 46422 slowly negotiates the level crossing at Glasson with a train of open wagons destined for the sidings alongside the wet dock. The brake van has been left alongside what remains of the station platform while shunting is carried out. The gate on the left is the entrance to the tiny goods yard. The date is the 2nd March 1962.

Ron Herbert.

Although business upriver at Lancaster was brisk, Glasson Dock itself appears to have been relatively quiet in the years leading up to the Great War. After an increase in activity following the introduction of the rebate for vessels carrying foreign cargoes, the timber trade appears to have diminished a little. In the middle of 1910, Huntingdon's sawmill at Glasson Dock was taken over by the Westmorland Cooperage and Sawmills Co. The firm had been established around 1901, at Holme, just over the county border in Westmorland, but had moved to Glasson Dock, bringing most of their workforce with them. Perhaps the convenience of easy access to both the dock and the railway was the deciding factor. The firm specialised in the manufacture of wooden boxes and packing cases and they also supplied timber cut to size for box manufacture. In their advertisements, they styled themselves as the Westmorland Sawmills Co but locally, the works was known simply as Read's, as it was owned by members of the Read family. Certainly, in 1912, the business had two partners, Wm Miles Read and James Fleming Read. In the years leading up to the Great War, Read's are known to have discharged several substantial cargoes of timber at Glasson Dock and may have used the canal basin as a timber pond for storing logs. The output from the works seems, from the outset, to have travelled to its destination by rail. This was certainly the case in the 1920s. Although Huntington's had relinquished the sawmill, they appear to have retained some sort of presence at Glasson Dock and continued to import significant amounts of timber and to store it on the quayside, alongside a firm from Carlisle by the name of J & JR Creighton. It is very likely that at least some of this activity generated traffic for the railway.

In February 1913, the LNWR wrote to the Commissioners suggesting that the coal tip be dismantled as it had not been used for some time, given that coal was now loaded onto vessels using the steam cranes equipped with their buckets. It will be recalled that, as part of the 1878 agreement between the railway company and the Commissioners, the former was to pay to the latter an annual sum equivalent to 4.5% of the capital cost of the sidings and appliances. Notwithstanding the use of the buckets and steam cranes, the Commissioners' register of shipping indicates that by this time, outgoing cargoes of coal were somewhat infrequent, although, of course, coal continued to be brought to the dock by rail for bunkering purposes. As a consequence of these changes, the railway company found itself in a position where it was paying an annual fee for a coal tip for which it no longer had any use. After some deliberation, the Commissioners replied, saying that although the tip had not been used for some years, it might be required in the future and consequently it should remain. They

Although this image is not of the best quality, it illustrates the footbridge that Williamson's erected across the branch at the rear of Lune Mills in 1903. This was to enable employees to safely access the site *via* the rear gate. Hitherto they had been obliged to dodge the trains to and from Glasson Dock and moving wagons being shunted on the quay goods line. The photographer is standing next to the main running line looking towards Lancaster.

CRA, Rev J Jackson Collection.

A fine view of *Engineer Lancaster* standing outside the LNWR's shed. The locomotive is a member of the Allen 6ft single class, dating from the 1850s but has been rebuilt with a large boiler and Webb cab and boiler fittings prior to being transferred to departmental use. Between 1891 and 1903, two different members of this class assumed the identity of *Engineer Lancaster*. However, in the absence of a date for this photograph it is not possible to say which of the two this particular locomotive is. Courtesy Lancaster City Museums.

The Lancaster District Engineer's inspection saloon, together with a party of gentlemen who have every appearance of being on some sort of day trip or outing. Unfortunately, the nature of the occasion is not recorded. When not in use, the saloon was kept in its own carriage shed at the Lune Road Engineer's Yard at the rear of Lune Mills. The date of the photograph and the precise location are unclear, but the mode of dress suggests that it was taken around 1910. L&NWRS.

Lancaster Castle Station looking south. Judging by how clean and new everything looks, this was taken very shortly after the completion of the rebuilding of the station in 1902. This work included the provision of a new Lancaster No 4 box situated to the north of the junction for the branch. The new twin-bay platform on the far right was used by the passenger trains to Glasson Dock as well as by services to the LNWR's Morecambe terminus at Euston Road. Lens of Sutton Association.

admitted that it was out of order at the present time and said that by way of a concession, they would not ask the LNWR to repair it. However, should it be required in the future they would ask the railway to put it back into working order.

Before going on to discuss the impact of the Great War on the branch, mention must be made of the establishment by the LNWR of a district engineer's workshop and depot on a site at the rear of Williamson's Lune Mills but on the opposite side of the railway. Pedestrian and vehicular access was from Lune Road. On 18th March 1904, the LNWR board read a minute from the Permanent Way Committee regarding the proposed transfer of their workshops and depot to the Glasson Dock branch. The estimate for the cost of the land required was £3,000, and for the new shops, machinery and sidings, £8,050. In January 1906, the board instructed that arrangements should be made with Lancaster Corporation for a supply of electricity to the site to work the machinery. The new works appears to have replaced an existing facility situated immediately to the south of Castle Station on the west side of the line. Rail access to the new site was from the main running line *via* a trailing connection for trains travelling towards Glasson Dock. The points for the new connection were worked from a two-lever ground frame, which was unlocked by a key in the end of the staff for the Glasson Dock branch. The new work was inspected by Colonel York of the Board of Trade and passed as satisfactory on 21st August 1905.

Initially, the Lune Road works and the yard seem to have been concerned only with permanent way maintenance and manufacture. However, later on, possibly after the Great War, it appears to have diversified to include other work, in particular, the manufacture of specialist timber items like replacement steps for signal boxes. In British Railways days, the site is said to have manufactured pre-fabricated wooden permanent way huts. In addition to these functions, the District Engineer's inspection saloon was kept at Lune Road in its own carriage shed, an arrangement dating from the establishment of the yard. For many years this was a purpose-built carriage with a veranda at each end and mounted on a 30ft 1in six-wheeled chassis. This was constructed in the half-year ended May 1891 and appears to have been allocated to the District Engineer at Lancaster from new. Around 1941, it was replaced with a modern purpose-built bogie carriage. It is recalled that, in British Railways days, this vehicle was fitted with a kitchen and this was the haunt of a steward who prepared and served meals when the saloon was in use. When not required for these duties, he filled in his time as a cleaner and maintenance hand in the yard. For many years, the District Engineer also had a dedicated engine for his own use. The locomotive carried cast plates bearing the legend *Engineer Lancaster*. There were in fact several iterations of *Engineer Lancaster* as it was the practice to transfer old locomotives to departmental use where they would be renamed and live out the remainder of their lives hauling the engineer's inspection saloon or working engineering or permanent way trains. The earliest of these was Allen 6ft single No 367 *Nightingale*, which assumed its new identity in 1891, before being scrapped in 1895. Its successor was No 110 *Canning*, another Allen 6ft single. This bore the title from 1895 until it was renamed *Engineer South Wales* and transferred away from Lancaster in 1903. The third *Engineer Lancaster* had formerly been Samson Class No 414 *Prospero*, which was built in 1873. This was allocated to the District Engineer until withdrawn in 1924. The final LNWR engine to bear the name was a member of Webb's 2-4-0 Waterloo Class, popularly known as Small Jumbos. This had previously been LNWR No 737 *Roberts* and was brought into departmental stock in February 1924. The locomotive appears to have remained in this guise until withdrawal in June 1935. All of these engines were attached to Lancaster's LNWR shed and, once the Lune Road yard was established, they would have been regular visitors to the Lancaster end of the branch. After the demise of the last *Engineer Lancaster*, any suitable locomotive was used but by the early 1960s this was usually an Ivatt 2-6-0. Engine No 46433 is remembered as one of the last steam locomotives to be used on the inspection duties. Originally, the crews for the inspection trains were sourced from Lancaster LNWR shed. However, when this closed in 1934, Lancaster's ex-Midland shed at Green Ayre provided both the locomotives and the crews. It is recalled that, in BR days, the Lune Road yard was also the home for the tunnel inspection train and for a petrol-driven trolley used by the permanent way maintenance gang on the branch.

Chapter Seven

The Great War and its Aftermath

This atmospheric scene taken in the early 1920s shows Glasson Dock with the photographer looking from the seaward end of the wet dock back towards the canal basin. The nearest vessel is the Fleetwood steam trawler FD152 *Davara*. It is likely that she is having some repairs carried out at Nicholson's shipyard out of sight on the right-hand side. Behind the trawler's smokestack are neat rows of the gas shells, which, in the aftermath of the Great War, were brought to Glasson Dock by rail to be dumped at sea.
Courtesy Lancaster City Museums.

WITH THE OUTBREAK of war with Germany on the 4th August 1914, Britain's railways came under government control on the following day. This control was exercised by way of the Railway Executive Committee whose task was, in essence, to co-ordinate the operation of all of the railway companies to ensure that collectively, they made the most efficient contribution to the nation's war effort. Over the duration of the conflict, the railways as a whole came under increasing pressure as industry gradually geared up to supplying the requirements of an all-out war of attrition carried out on an unprecedented scale. The conflict was to affect the branch in a number of ways. However, as always, the goods traffic was significantly dependant on ships arriving at either Glasson Dock or upstream at Lancaster and also on the activities of Williamson's, therefore, it is necessary to step back a little from the railway itself and look closely at these peripheral aspects in order to better understand the effect of the conflict on the branch. As might be expected, the war had a significant and cumulative effect on the shipping industry. Following the outbreak of hostilities, the threat of enemy action limited the use of ports on the east coast and the Royal Navy requisitioned areas at a number of dock facilities. Increasingly, various segments of the shipping freight market became regulated and as hostilities continued, an increasing number of merchant ships were requisitioned, or diverted to war work. All of this caused considerable disruption to existing commercial traffic and, above all, pushed up shipping freight rates to the point where they were no longer competitive. The situation was exacerbated from 1916 onwards as the Germans intensified their U-Boat campaign, especially after February 1917 when a switch was made to unrestricted submarine warfare. At Glasson Dock itself, the cargoes of wood pulp for Cropper's ceased with the outbreak of war, although the timber trade continued for a little while longer. It is likely that most of this timber would have reached its eventual destination by rail. During 1915, cargoes of china clay were still being brought to Glasson Dock for both Cropper's and Storey's. Shipments for the former would certainly have been taken by rail to Burneside. There were also still a number of instances of cargoes for Williamson's being discharged at Glasson rather than at the New or Ford quays. In each instance this seems to have been because the vessel in question was just a little too large to reach Lancaster in safety. It is possible that some of these cargoes were discharged into railway wagons. However, by this time, Williamson's owned a small fleet of motor lorries and the use of this mode of transport for the final part of the journey up to Lancaster cannot be discounted.

In 1916, there were two small cargoes of timber for Huntington's and a larger one for Read's. Carnforth Iron Works brought two small ships to Glasson Dock laden with iron ore, which must have brought back memories of busier times at the port. Unfortunately, these were not repeated. By 1916, a new ship owner appeared on the scene, Harry Dingle, whose vessels began making regular (and no doubt hazardous) voyages from

Taken at Glasson Dock in the early 1920s, this shows a Furness Railway six-wheeled wagon carrying what appears to be one of the heavy surf boats built for the Elder Dempster Lines by Nicholson's shipyard. Footage of this type of boat in action at Accra in Ghana in 1947 is currently available on YouTube.

Author's collection.

Glasson Dock, with coal destined for Gibraltar, Ireland and Spain. This would have been brought to the port by rail and loaded by the steam cranes. Dingle also brought regular cargoes of macadam to Glasson, as well as a significant cargo of timber and a large cargo of animal horns. It is likely that at least some of these commodities continued their journeys by rail.

By the year 1917, traffic to the port as a whole had fallen off catastrophically. Upriver at Lancaster, cargoes to Williamson's had now reduced to a trickle. At least some of this reduction probably reflects a move to rail transport owing to the difficulties within the shipping industry during the war years. However, as the conflict wore on, there is no doubt that there was a significant curtailment of the output from Lune Mills and St George's Works. As a firm that exported a substantial portion of its products abroad, Williamson's was especially vulnerable to the effects of the shortage of shipping space and the rising cost of transport by sea. Furthermore, as the war progressed, it became increasingly difficult to source raw materials especially those coming from abroad. To make matters worse, towards the end of 1917 there was a severe shortage of linseed oil and, after meeting the government's priority requirements, none could be spared for the linoleum industry. Finally, Williamson's with its large, unskilled workforce, was severely affected by losses of manpower due to employees volunteering for the armed forces and, later on, being conscripted. By 1917, production at Lune Mills was being sustained in only two-thirds of the works, with the size of the workforce at less than a half of its pre-war level. Unsurprisingly, commercial traffic at Glasson Dock was now also somewhat limited. However, the outlook was not entirely gloomy. Nicholson's shipyard, situated on the south side of the wet dock, was very busy indeed.

The shipyard at Glasson Dock, so far, has only been touched upon briefly, therefore, a *résumé* of the history of both the site and the firm of Nicholson's is appropriate at this point. In 1841, the Port Commissioners had opened a graving dock at Glasson, access to which was gained through the wet dock. After initially running the facility themselves, the whole site was leased to a succession of tenants. These operated a ship repair business but also built new vessels in a small shipyard established alongside the graving dock. At the time the branch line was opened in 1883, the tenants were Richard Nicholson and William Marsh. The death of the latter in 1896 resulted in the dissolution of the existing partnership and the creation of a new one, styled Nicholson and Sons. In the period 1882 to 1906, 13 wooden vessels of different tonnages were built at the yard. However, ship repairing remained the staple of the business. Over the years, Nicholson's also carried out a great deal of day-to-day work for the Port Commissioners, constructing buoys, maintaining the Commissioners' tug and dredger and carrying out maintenance and repairs on the dock estate. After 1904, there was some extension of the premises when the firm at first rented, and then later acquired, fields to the west of the shipyard and erected more buildings. It seems that from the beginning of the war, Nicholson's ship-repair business expanded dramatically. Local memory indicates that during the conflict, a significant proportion of the space in the wet dock was taken up with vessels either being worked upon or awaiting attention. It is recalled that much of this work was carried out on steam trawlers from nearby Fleetwood. The shipyard was never actually connected to the railway, being on the opposite side of the wet dock from the quay lines. However, the expansion of business was such that towards the end of the war, plans were made to extend the shipyard's facilities and to construct a siding to it. The business must always have generated some traffic for the LNWR and this is likely to have taken the form of coal, coke, iron and steel in various forms, together with engine and boiler fittings. The shipyard also had its own sawmill and it is likely that, from time to time, specialist timbers were brought in by rail. During the war years, this rail traffic must have increased significantly in line with the additional business carried on by Nicholson's.

Over the 16th–17th May 1917, the branch played host to the royal train, which was stabled overnight at Glasson Dock. On the 16th, the train arrived at Lancaster, having travelled from Manchester, and the King and Queen alighted at Castle Station to visit the munitions factories at White Lund near Morecambe and on Caton Road on the east side of Lancaster. With the royal couple away on official duties, the train was taken down to Glasson Dock and stabled overlooking the estuary. Local memory recalls that the train was taken down the branch by two 0-6-0 tank engines from the LNWR's Lancaster shed. These were specially cleaned for the occasion. However, there were no cleaners at Lancaster owing to wartime manpower shortages and consequently two men had to be borrowed from Preston shed to carry out the work.

After visiting the two factories, the King and Queen travelled to Glasson Dock by motor car and spent the night on the train. An engine in steam remained coupled to the carriages throughout their time at Glasson Dock, in order to provide heating. It is recalled that during the night, an urgent message had to be sent up to Lancaster requesting a second locomotive, as the first one was running low on coal and water. The following day, the train departed from Glasson Dock to Workington, travelling along the Furness Railway as far as Whitehaven. The official instructions covering the journey from Lancaster, as far as Furness Railway metals at Carnforth, have survived and provide some interesting details. It is clear that absolutely nothing was left to chance to ensure that the royal train had a safe and smooth passage along the railway. A pilot engine was to run 15 minutes in advance of the royal train and a distinctive block bell code "is the line clear?" consisting of 12 strokes split into three groups of four, was to be used for signalling the pilot engine and the train. The carriages were to be thoroughly examined before the start of the journey and the train was to be accompanied by a complement of experienced Carriage Department artificers who were to keep watch on either side of the train whilst it was travelling. In addition to this, men from the Telegraph Department were to accompany the train with the necessary apparatus to enable them to tap into the telegraph wires to notify the authorities in the event of an incident or accident. All facing points on the route were to be padlocked by the permanent way staff and remain so until after the royal train had passed. In addition to this, gates at unmanned level crossings were to be locked for at least an hour before the passage of the pilot engine. Platelayers were to be on duty on their respective lengths of track and were instructed to exhibit a green hand signal upon the approach of the royal train, in order to confirm that it was safe to proceed. Station masters were to be present on the station platform to observe the passing of the train and were responsible for ensuring that the railway employees under their control knew exactly what to do and performed effectively. The overall impression is of everyone working at concert pitch while the royal train occupied their area. The event must surely have placed a considerable amount of responsibility on the shoulders of the station master at the normally quiet backwater of Glasson Dock. No doubt there was a sigh of relief when the whole thing passed off satisfactorily and the train became someone else's responsibility.

As regards the specific arrangements for the outward journey from Glasson Dock, the two engines to work the train and the pilot engine were to leave Lancaster at 7-45 am and to travel tender-first down the branch, with the pilot engine at the rear. The latter was to leave Glasson Dock at 8-15 and having travelled up to Lancaster, proceeded to the shed, its work done. The royal train was to leave 15 minutes later. The special notice tells us that it departed from the harbour line, which makes it difficult to decide whether it was stabled alongside the station platform or on the line to the river quay. However, local memory once again comes into play and apparently a special platform was constructed for the occasion. This suggests that the train was stabled on the latter line. Once the royal train arrived in Lancaster station it was met by the pilot engine and two train engines provided by the Furness Railway, which would take it northwards as far as Whitehaven. The LNWR provided two guards who took charge of the train between Glasson Dock and Lancaster and again from Whitehaven to Workington. For travelling between Lancaster and Carnforth and then over their own lines, the Furness provided their own guard who joined the royal train at Lancaster. A small number of trains on the LNWR main line were held at signals or their departure was delayed so as to keep Castle station free while the engines were exchanged on the royal train. On the Glasson Dock branch, the 8-00 am train from Lancaster to the terminus and the corresponding return service at 8-35 were cancelled.

Notwithstanding the successful ship repair business carried on by Nicholson's, by 1918, normal trade at Glasson Dock and at the quays up river in Lancaster was virtually at a standstill. There were a couple of cargoes for Williamson's, one for Storey's and a handful of outgoing vessels loaded with coal going to Ireland and France. The Commissioners' dredger had been laid up in the river at Lancaster for the duration of the war and their steam tug *John O'Gaunt* had spent much of the period on loan to the government. Both of these vessels were disposed of in the 1920s and were not replaced.

The war finally ended on 11th November 1918. The conflict had significantly changed the political and economic landscape of Great Britain, and it would take some time before things settled into what was to be a new version of normality. The relatively settled patterns of trade and commerce, which had existed in the pre-war era, had gone for ever. The first full year of peace did not see any sign of a return to pre-war traffic levels. For the moment, Nicholson's shipyard remained as busy as ever, but the pattern of very limited commercial traffic remained the same as in 1918. However, by 1921, there were some signs of a revival of trade, mainly in the form of cargoes of maize, peas and beans for Lancaster. There were some cargoes for Williamson's and regular cargoes of macadam were being discharged at Glasson Dock. It is possible that some of this travelled inland by rail. However, at this point, foreign trade had not resumed and the amount of railway traffic generated by vessels calling at Glasson Dock during this period is likely to have been negligible. There was something of a change in the Spring of 1922 when HM Government's Disposal and Liquidation Board approached the Commissioners with a view to storing gas shells on the quayside at Glasson Dock prior to them being disposed of at sea. Initially the Commissioners refused, saying it would interfere with other traffic. However, it is difficult to understand exactly what traffic they had in mind, as Glasson Dock was somewhat quiet at this time. Perhaps it was simply an excuse to avoid dealing with this rather dangerous cargo. However, after an interview with representatives from the Board, they agreed to try one cargo as an experiment, subject to the Board providing an indemnity in the event of any accident taking place. The LNWR initially seem to have been similarly reluctant to have anything to do with the gas shells. The Commissioners received two letters from the railway company, the first one objecting to the use of the cranes on the docks for loading the shells, and then another one, two days later, agreeing to this. Perhaps they had also had an interview with the men from the Disposals Board. The shells themselves came from National Filling Factory No 13 at White Lund between Morecambe and Lancaster. This was a purpose-built facility and was served by extensive sidings coming off the Midland Railway's line to Morecambe. Local

Taken from the northern end of Lancaster Castle Station, the ex-LYR 0-6-2 tank is taking the Glasson Dock branch road with what is probably a train of coal wagons. The brake handle on the wagon has been pinned down and this probably applies to most if not all of the wagons in the train. This is to assist in keeping the loose coupled train under control as it cautiously makes its way down the curving 1 in 50 gradient.

AP Herbert, courtesy Ron Herbert.

This image appears to capture the short period in early 1932 when Williamson's began bringing all their coal supplies into Glasson Dock but before they installed two grab cranes running on a dedicated 7ft-gauge line. There are several piles of coal close to the pier head and an enclosure of railway sleepers has been constructed in which to store discharged coal. Almost all of this was taken by road to Lune Mills.
Commercial postcard, author's collection.

memory has it that the gas shells were transferred from White Lund to Glasson Dock by rail and then stored on the quayside until they could be taken out by ship and disposed of at sea. The initial shipment must have been deemed a success, as very quickly after this, there were regular cargoes of gas shells being taken out from Glasson Dock on the appropriately named SS *Depositor*. However, by December of 1922, there were complaints from various quarters about leakage from the shells while they were in transit to Glasson Dock or whilst being stored on the quayside. The Commissioners wrote to Dr Hampson, the Medical Officer at White Lund, saying that if any more complaints of this nature were received, the harbour master had been instructed to halt the delivery of the poison gas shells to the dock. There then followed a flurry of correspondence between the Commissioners, Dr Hampson and the Rural District Council, who also weighed in on the issue. In the end, it all seems to have been sorted out and the traffic continued throughout 1923, finally petering out in the early part of 1924. During the course of loading the SS *Depositor*, a small number of the gas shells had fallen into the dock and these could only be recovered in June 1923 when the dock was drained and cleaned for the first time since the war. It appears that some of the cargoes disposed of at sea consisted of cylinders of poison gas rather than shells. In May 1924, a "poison gas cylinder" had floated up the Lune and stranded close to Lord Ashton's golf house. At the same time, the harbour master at Barrow-in-Furness wrote to the Commissioners, reporting sightings of floating cylinders off Walney Island. All of these incidents were reported to the Disposals Board. It must have been something of a relief to all concerned when this rather dangerous traffic finally came to an end. For reasons which remain unclear, the Disposals Board was not charged for storage space at Glasson Dock. However, following the termination of the traffic, the Commissioners received a grant from the Treasury of £600 "in respect of the use of the Quays for the disposal of contaminated material".

When the Great War finally ended in November 1918, Britain's railways were in a rundown condition. The unprecedented levels of traffic generated by the war effort, in theory at least, should have provided a significant increase in income for the privately owned railway companies. However, once under Government control, their profits had been restricted to their pre-war level with the financial results for 1913 being used as a benchmark. Any surplus income went into the Treasury's coffers to assist with the war effort. As the conflict wore on, owing to shortages of manpower and essential raw materials and stringent Government controls, it became increasingly difficult to procure new locomotives or even to carry out essential repairs. One of the consequences of these unusual and difficult operating conditions was that by the end of the conflict, unprecedented arrears of maintenance had accumulated. Furthermore, the wartime limitations on profits had left some of the railway companies close to bankruptcy and no longer in a position to help themselves. The railways, quite correctly, were perceived as being an important factor in the nation's post-war economic recovery and there was a feeling in Government circles that some sort of fresh start was required. In August 1919, the Ministry of Transport was formed, and one of the tasks of the first Minister of Transport, Sir Eric Geddes, was to plan for the future of the railways in Great Britain. In the period July 1920 to May 1921, the Ministry entered into wide-ranging consultation with interested parties. The final outcome was the decision to merge the existing railways into four large regional companies. These were to be the Southern Railway, the Great Western Railway, the London and North Eastern Railway and the London Midland and Scottish Railway. These mergers were to take effect from 1st January 1923. These far-reaching provisions came together in the 1921 Railway Act, which received royal assent on 19th August 1921, just four days after the wartime government control of the railways finally ended on the 15th of August. Thus, on 1st January 1923, the Glasson Dock branch became a part of the newly formed London Midland and Scottish Railway.

In the years following the Great War, the Lancaster end of the branch seems to have very gradually recovered much of its goods traffic, owing principally to the activities of Williamson's and the Corporation gasworks. The same cannot be said for the Glasson Dock end however, which appears to have remained relatively quiet. In terms of shipping, by 1928, there was no foreign trade at the dock itself and cargoes were limited to regular shipments of macadam, occasional cargoes of china clay and some consignments of oats for W & J Pye. By

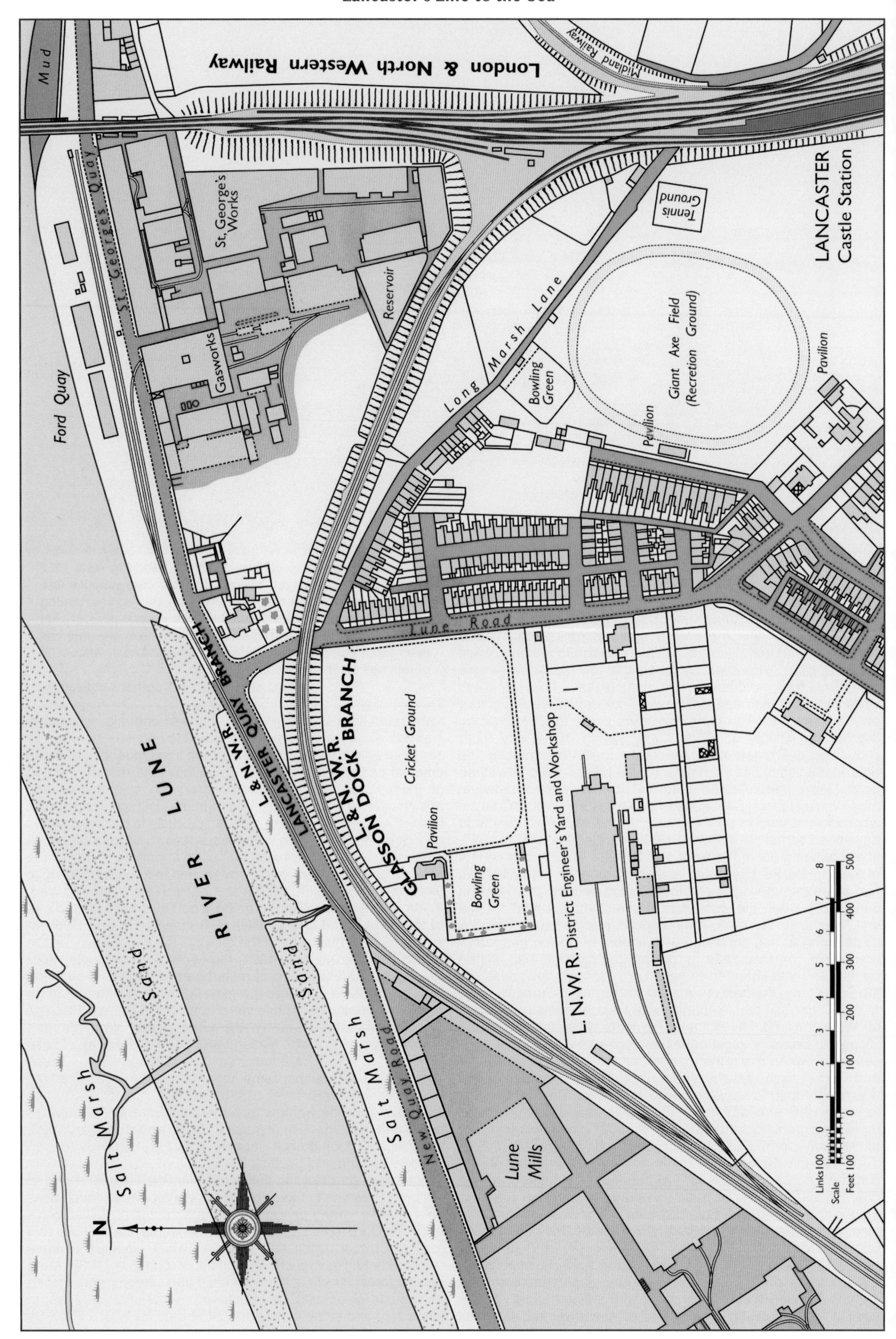
London & North Western Railway
Midland Railway
Mud
St. George's Quay
St. George's Works
Gasworks
Reservoir
Ford Quay
Long Marsh Lane
Tennis Ground
LANCASTER
Castle Station
Giant Axe Field (Recretion Ground)
Pavilion
Bowling Green
Pavilion
Lune Road
L. & N. W. R. LANCASTER QUAY BRANCH
L. & N. W. R. GLASSON DOCK BRANCH
Cricket Ground
Pavilion
Bowling Green
L. N. W. R. District Engineer's Yard and Workshop
RIVER LUNE
Sand
Sand
Salt Marsh
Salt Marsh
New Quay Road
Lune Mills
N
Links 100
Scale
Feet 100

The Lancaster end of the branch on the eve of the Great War. The Ford Quay complex with Williamson's St George's Works and the Corporation gas works can be seen top right. Lower left of centre is the LNWR District Engineer's yard. Opposite the yard can be seen the crossover connecting the main line to Glasson Dock with the quay goods line. This was installed in 1910 to facilitate shunting movements in the area. Finally, a small part of the sprawling Lune Mills site intrudes from the left
Map redrawn by Alan Johnstone.

this date however, it is quite likely that at least some of these commodities were leaving Glasson Dock by road rather than by rail. Nevertheless, the industries clustered around the village continued to generate some traffic for the branch and we will examine this in detail in due course.

Although the tonnage figures for the Port of Lancaster as a whole had dipped significantly in 1924, owing to the cessation of the gas shell disposal contract, the 1920s saw a very gradual recovery of at least some of the peacetime shipping traffic. However, the annual figure of net registered tonnage, stubbornly refused to climb much above 10,000, about a third of the pre-war figure. A new shipowner who quickly established himself at the port was Robert Gardner. In 1920, he had taken over the small shipping business of Harry Dingle and from that point, gradually acquired a fleet of small coasting vessels. One of his earliest ventures was the disposal of the gas shells from Glasson Dock and it was he who either owned or chartered the SS *Depositor.* Right into the 1950s, Gardner, was a frequent user of the port facilities at both Glasson Dock and Lancaster. In addition to his activities as a shipowner, he also ran separate businesses as a builder's merchant and as a coal merchant. The latter was operated from sidings at the Midland's Green Ayre station and was established there by 1922. By the 1930s and possibly sooner, Gardner is known to have operated privately-owned coal wagons with his name emblazoned on the side. As well as supplying house coals and slack, Gardner advertised anthracite and different grades of patent smokeless fuel, which seem to have been brought in by sea from South Wales. It is possible, therefore, that his railway wagons were to be seen on the Ford Quay, loading coal from one of his coasting vessels.

During 1920 and the early part of 1921, there had been protracted negotiations between the Commissioners and T W Ward, of Sheffield, who wished to use the river quay at Glasson dock for ship breaking and for the laying-up of ships. Similarly, in 1923, the Anglo-American Oil Co approached the Commissioners with a view to establishing an oil storage facility at Glasson. Plans were drawn up and a group of Commissioners visited the Anglo-American Oil Co's facility at Barrow-in-Furness to gain an understanding of what was required. Local memory recalls that, in July 1924, the LMS laid on a special train to take representatives from the Port Commission, the railway and the oil company to Glasson Dock for an on-site planning meeting. Both Ward's and Anglo-American would have brought additional business to the port and to the railway, but unfortunately these two projects were not proceeded with. With the failure of these two schemes, it seems to have become apparent that some sort of fresh initiative was required. Glasson Dock now saw little use and the facilities there, and at the quays and berths at Lancaster, were suffering from the arrears of maintenance that had accumulated during the war years. Pre-war traffic levels had simply not returned and consequently, the income from the port was significantly reduced. All interested parties agreed that something needed be done, but the Commissioners simply did not have the funds to carry out the work necessary to make the port more attractive to shipping. In 1923, they had managed to find sufficient money to replace the outer dock gates at Glasson, but whilst constituting an essential item of maintenance, this did nothing to enhance the facilities at the port. In May 1926, there was a meeting between the Commissioners, representatives of the LMS and the shipowners to discuss the possibility of the port being taken over by Lancaster Corporation. The shipowners wanted improvements and investment, of course, and whilst the Commissioners, agreed in principle to a takeover, they were very doubtful that the Corporation could be persuaded to take on the financial burden. In the end this proved to be the case. For its part, the newly-formed LMS was now making ominous noises about the branch not being remunerative. Under the terms of the 1878 agreement and the Act of Parliament authorising the construction of the branch, the railway company was responsible for the maintenance of the aging steam cranes and the redundant coal tip at the dock. In addition to this, they were paying a fixed rent of 4.5% *per annum* on the capital cost of these and on the quayside sidings owned by the Commissioners but worked exclusively by the railway. These were now used very infrequently and, in the increasingly bleak economic climate of the 1920s, the railway company's sole concern, quite understandably, was to cut expenditure on the branch and its infrastructure to a minimum.

In spite of this rather grey outlook, there was still some railway goods traffic both to and from the station at Glasson Dock. In the Bowtell Collection held at Kendal Archive Centre, there is a

A view at the rear of the Lune Mills site looking up the branch towards Lancaster. The track on the left is the quay goods line, whilst the one on the right is the route to Glasson Dock. The gates into the district engineer's yard are open. The building behind the fence with the curved roof is the carriage shed used to house the engineer's saloon carriage. On the left is the short siding used to discharge tank wagons.

CRA, Rev J Jackson Collection.

copy of a letter from Mr E W Gilmour, who was station master at Glasson Dock in the period 1927–1930. In it he describes the branch as it was during his time in charge. There also exists a copy of a longer letter from Mr Gilmour, in the possession of retired railwayman and author Ken Nuttall of Lancaster. Taken together, the two documents provide a detailed portrait of the Glasson Dock end of the branch as it was during the late 1920s and we will return to these recollections at various points in the narrative. In relation to the goods traffic at Glasson Dock, the former station master describes the principal outbound and inbound traffics as follows:

Outbound

1) *Mussels, not fit for human consumption but sent to various east coast ports for bait. This amounted to one or two wagons daily in season.*
2) *Occasional loads of draining tiles carted by a horse and cart from the "brick croft" situated just outside the village.*
3) *Cargoes of china clay for Storey's of Lancaster. This was brought by sea from Cornwall and transferred into railway wagons. A shipload of about 400-500 tons arrived at the port three or four times a year.*
4) *Boxwood and wooden boxes from the sawmills This was loaded into vans from the firm's premises on the side of the canal basin. This was the most regular and valuable outward traffic and usually amounted to two to three wagons daily.*

Finally, although not strictly goods traffic, Mr Gilmour recalls that from April to August, salmon were caught in the River Lune by local fishermen and dispatched by rail. In the season, this constituted the biggest single item in the station's parcel account. During Mr Gilmour's tenure at the station there was also one farmer who sent his milk away by train on a daily basis.

Inbound

1) *A seasonal traffic of timber for the sawmill. This was discharged at Barrow-in-Furness and came from various Baltic ports. This arrived during the Autumn months and during this period there might be up to five or six wagons daily.*
2) *Oats and other animal foods for the corn mill*
3) *Steelwork for the shipyard*
4) *Coal for local use* and for bunkering ships

Some commentary on all this is appropriate. The traffic in mussels appears to have been a well-established one and there is a reference to it in a local newspaper in 1917. Towards the end of Mr Gilmour's time as station master, purification tanks were installed at Glasson Dock and the mussels, now fit for human consumption were sent by passenger train to various Lancashire towns. The "brick croft" was operated by Messrs J Robinson and Sons. The firm seems to have specialised in drain tiles and, as early as 1890, their newspaper advertisements were offering these for delivery at any railway station or any wharf on the canal. The firm certainly used the railway to transport their finished products. For example, it is known that in 1913, a load of 11 tons of "tiles" was transported by rail from Glasson Dock to Sandside on the Furness Railway. These were presumably drainage tiles and could only have come from Robinson's. Local memory has it that in the earlier years of the twentieth century, the coal for the kilns arrived by canal rather than by rail. This makes perfect sense, as the site was much closer to the former than to the railway station. As regards the china clay, there is evidence that, even in the late 1920s, cargoes for both Storey's and Williamson's were occasionally discharged at Glasson Dock. The infrequency of the shipments suggests that these were instances where the vessel in question was too large to navigate the river channel up to Lancaster. The sawmill, of course, was operated by the Westmorland Cooperage and Sawmills Co. Foreign cargoes of timber to Glasson Dock seem to have ceased in 1923 and it appears that from then on, the firm's requirements were brought in by rail. It is likely that this would also have been the case for any timber required by Nicholson's shipyard.

The corn mill referred to in Mr Gilmour's list is the one belonging to Pye's at Thurnham, just a short distance up the Lancaster Canal from Glasson Dock. During the harvest season, one or two wagon loads of oats were received for the firm each day. These came from stations in the Ormskirk and Southport areas. During the late 1920s, Pye's were also still bringing regular cargoes of maize and oats up the river to the Ford Quay, where they continued to rent one of the Commissioners' sheds. The firm also brought occasional ship loads of oats and basic slag (a phosphatic fertiliser) into Glasson Dock. Pye's was a significant concern, with a number of mills and warehouses in the wider Lancaster area. For many years they had a presence in the goods yard at Pilling station on the Knott End branch. However, in 1923, the firm constructed a new warehouse immediately to the west of the station, which was served by a private siding. For transport purposes, Pye's used a mix of coasting vessels, the railway and motor lorries. A reminder that in the 1920s and 30s, the traditional model of a straight switch from rail to road carriage does not always reflect what happened in practice, with some firms opting for a tailored solution best suited to their business needs.

The shipyard mentioned by Mr Gilmour, of course, is that of Messrs Nicholson. By 1928, the wartime boom in shipping repairs had long since come to an end, but business appears to have remained steady. In the early 1920s, the company had obtained a contract to construct a number of wooden surf boats for the Elder Dempster lines. These were sturdy rowing boats of a particular design which were used on certain parts of the West African coast to take heavy casks or sacks of produce out to large ocean-going vessels waiting offshore. At the time of writing there is some excellent footage on YouTube showing this type of boat in action at Accra, Ghana in 1947. There is photographic evidence to show that at least some of the completed boats left Glasson Dock by rail.

As regards coal being brought to Glasson Dock, we have already seen that the local tile works of J Robinson and Sons obtained their supplies of coal using the canal, and carriage by canal is a possibility for local requirements even in the 1920s. For many years, coal from the Wigan area was taken by rail to the terminus of the Lancaster Canal in Preston. Here, it was tipped into barges and taken north to various destinations. The prominent Lancaster coal factors, Samuel Thompson & Co. are known to have operated their own barges on the canal as well as owning a fleet of railway wagons. This was a not uncommon arrangement on the Lancaster Canal and, indeed, on the Leeds and Liverpool Canal further south. The pre-railway phase of industrial development had stimulated the construction of mills and other industrial premises right alongside the canal for the efficient reception of raw materials and for the transport of finished goods. Even after the establishment of a comprehensive railway system, it was still more convenient and economic for businesses using these premises, to receive their coal and raw materials by water. Therefore, well into the twentieth century, as well as Thompson's, WJ Turner (of Wigan) and the Wigan Coal & Iron Co, both operated barges on the Lancaster Canal in addition to owning their own railway wagons. Incidentally, the Port Commissioners themselves were very long-standing customers of Thompson's. From the opening of the branch line, the firm were all but exclusive suppliers of steam coal for the Commissioners' tug and dredger and also house coal for heating the Commissioners' premises. This was always ordered by the wagon-load and up to three wagons might be delivered to either Glasson Dock or to the quay at Lancaster, whichever happened to be the more convenient. We have already seen that Thompson's were also involved in the export of coal from Glasson Dock, and they are also known to have supplied coal to the Corporation gasworks. Over the years, their wagons must have been a common sight on the branch. The Commissioners' vessels, of course, were disposed of in the early 1920s and not replaced. However, at the beginning of the Second World War, the firm was still supplying household coal to the Commissioners' Cockersands Abbey lighthouse near Cockerham. As a point of interest, for a number of years, the lighthouse was manned by what must have been one of Britain's few female lighthouse keepers, Miss Janet Raby, who took over the role following the death of her brother, the previous holder of the post.

Any steam ship ending its voyage at Glasson Dock or, indeed, at the quays at Lancaster, is likely to have had to replenish her bunkers before going to sea again. There is evidence that during the Great War and afterwards, Nicholson's shipyard stored

Wagons belonging to the Lancaster coal factor Samuel Thompson & Co would have been a common sight on the Glasson Dock branch. For many years the firm had establishments in Lancaster, Preston and Kendal and it appears that each had its own wagons, lettered accordingly. In the absence of a good picture of a Lancaster-based wagon, this view of one from Kendal will have to suffice. Simply replace the word "Kendal" with "Lancaster". These wagons are said to have been painted red with black ironwork and white lettering shaded black. The motif was a white disc edged in black with a red cross.

CRA, PO Collection.

quantities of coal on the quayside for bunkering newly-repaired vessels. Mr Gilmour, the former station master, estimated that the railway brought in around 60 tons of coal each month in order to keep these stocks of bunkering coal replenished. Given the quantities involved, individual ship owners are more likely have dealt directly with the colliery or with one of the larger coal factors. If the ship owner was based in a port some distance away, and had a regular contract with a particular supplier, coal wagons from collieries some distance from the port might suddenly put in an appearance at Glasson Dock. It is known that as late as 1932, enquiries were raised with Dinnington Main and Rossington Main collieries (both in South Yorkshire) concerning rates charged for supplying bunkering coal at Glasson Dock. There is photographic evidence of the presence on the quayside of wagons from Bickershaw Colliery near Leigh in south Lancashire and from New Ingleton Colliery at Ingleton, just over the border in Yorkshire. The latter was the nearest source of coal to Glasson Dock. It is likely that both concerns were supplying coal for ships' bunkers. Also visible in the background of photographs of the quayside in the early years of the twentieth century are wagons from the Wigan Coal & Iron Co, Pearson and Knowles Coal and Iron Co, the Moss Hall Coal Company (both from the Wigan area) and finally Park Lane Colliery, which was situated at Bryn between Ashton and Wigan. There is further evidence relating to the coal traffic on the branch in an image taken in the 1920s of a goods train leaving Lancaster Castle station and taking the Glasson Dock line. The first wagon belongs to W H Shaw, a firm of coal factors based in Haslingden, East Lancashire. The remainder of the train is unfortunately not visible, but the appearance of a factor's wagon from outside the area suggests that this might have been part of a bulk purchase for bunkering purposes. Alternatively, it could have been for a significant customer on the branch, the Corporation gas works on the Ford Quay or Williamson's being the most likely recipients.

In the late 1920s, although the goods traffic on the branch was holding up satisfactorily, especially at the much busier Lancaster end, the same could not be said for the passenger traffic. In May 1929, the Commissioners received a letter from the railway company to the effect that they were considering closing the Glasson Dock line to passengers and that they would be in touch once something more definite was known. At this time, the passenger services on a number of local branch lines in the Lancaster–Preston area, were facing severe competition from cheaper and more frequent motor bus services. However, in strictness, this was not the case at Glasson Dock, as there was no service into the village itself. Buses to and from Lancaster could only be accessed at Conder Green. As the crow flies, this was about half a mile from the village. However, the river Conder and the surrounding marsh provided significant obstacles and so, to reach Conder Green on foot, meant a journey of perhaps just over a mile by road, across open and exposed countryside. As we have seen, in May 1926, during the course of the meeting with the Commissioners, the LMS had made it clear that the branch, as a whole, was not paying its way. In spite of this, or perhaps because of it, the railway company does appear to have made an effort to provide a service that would be attractive to the travelling public. Certainly, the passenger timetables for the 1920s bear this out. For the earlier part of the decade there were daily five trains in each direction and additional trains on Wednesdays and Saturdays, which were market days in Lancaster. However, by 1929, this had reduced to four daily trains and the Wednesday-only trains had disappeared. Writing about his experiences as the last station master at Glasson Dock, E W Gilmour is quite disparaging about the viability of the passenger service in the later 1920s, averring that if the booking office took £1 across the four daily trains with £3 on Saturdays, they were considered to be doing well. He also goes on to describe how, on occasions, the passenger receipts received a welcome boost. When ships came to Nicholson's shipyard for repairs, it was the usual practice for the shipowners to pay off all or most of the crew members. When this happened, a group of perhaps six to ten men would make their way along the road to the station and book single tickets to their homes. Mr Gilmour recalls Whitehaven, Liverpool, Preston and Cardiff as destinations. Similarly, it appears that in the 1920s, the shipyard did not employ boiler makers on a permanent basis, so from time to time they would bring men in from Birkenhead and Barrow. They would work temporarily at Glasson Dock for perhaps several weeks at a time but would travel home by rail at weekends. These longer distance fares provided a welcome increase to the rather meagre passenger receipts.

In February 1930, representatives from the Port Commissioners were invited to a meeting in the offices at Lancaster Castle Station to discuss the proposed closure. The LMS explained the legal position and pointed out the heavy loss incurred on the service, which had amounted to £1,500 in the last financial year. They proposed replacing the passenger trains with buses and promised to provide a service as good as, if not better, than currently available on the railway. This was certainly within the power of the LMS as, in 1929, following the Railways (Road Transport Act) of the previous year, the company had taken the opportunity to acquire a significant shareholding in Ribble Motor Services Ltd of Preston. By 1930, this company had acquired a virtual monopoly on the bus routes between Lancaster and Preston and Blackpool. The Commissioners wanted the service to go as far as the Victoria Hotel in the centre of the village, rather than terminate at the railway station, and also for the buses to convey small parcels. The LMS appears to have been agreeable to this and, therefore, the Commissioners gave their assent to the cessation of the passenger service. Perhaps they were slightly relieved that there was no attempt to close the line completely, which, at this time, would have had

One of the batch of six tank wagons constructed in 1931 for Williamson's by Charles Roberts & Co, the Wakefield-based wagon builders. It is unclear what these were used for, but the likelihood is that they were put to work carrying linseed oil, which was used in large quantities in the manufacture of linoleum. At some point, probably in the 1930s, a short siding was constructed at the rear of Lune Mills to be utilised for the discharge of tank wagons. CRA, PO Collection.

a significant effect on the already limited viability of the port facilities. A report on the forthcoming closure appeared in the *Lancashire Evening Post* for 9th May 1930. The formal public notice of closure appeared in the local newspapers a little later. This was to take place on and from the 7th July, and the stations at Ashton Hall Halt, Conder Green Halt and Glasson Dock would be closed to passengers. This date was a Monday, and so, as there was no Sunday service, the final passenger trains ran on Saturday the 5th. Ribble Motors would operate a motor omnibus service between Glasson Dock and Lancaster Castle Station and a separate announcement would be made in relation to this. The notice reassured readers that holders of railway season tickets would be able to travel on the new buses until their tickets expired, as would holders of return halves of railway tickets issued prior to 7th July. Glasson Dock station would continue to deal with parcels and miscellaneous traffics as well as horses, cattle and other livestock. With the passenger service ended, the LMS was anxious to obtain some income from the vacated passenger accommodation at Glasson Dock and, in September 1930, this was advertised to let, along with similar premises on the Knott End and Longridge branches, both of which had also lost their passengers services that year.

In May 1930, Lord Ashton died aged 87. He had left his imprint on Lancaster in so many different ways, not least because his linoleum and floorcloth empire had been and still remained, the largest employer in Lancaster. There was speculation in the town that the LMS had not dared to withdraw the passenger service from Ashton Hall Halt until his death. However, there is no evidence to support this. Certainly, the LMS seem to have decided to close the line to passengers well before the aged Lord shuffled off his mortal coil. Furthermore, there is no evidence that the halt was ever actually used by Ashton, although it may have been used by the workers on his estate. From around 1920, most of the passenger trains had third class accommodation only, surely a disincentive for a peer of the realm. Whether true or not, the story provides an interesting example of the depth of James Williamson's influence on the economy of the town and on the minds of its inhabitants. Remarkably, to the end, he had run the business single handed, having no partners and never succumbing to the lure of creating a public or even a private limited company. He did not leave a will and on his death, he left an estate with a preliminary valuation of around nine and a half million pounds, a record for intestacy. In August 1931, the business was reconstituted as a private limited company under the chairmanship of Williamson's son-in-law, the first Earl Peel. Inevitably, this new era ushered in a gradual modernisation of the works, as well as some changes in the way the business was run. Throughout Lord Ashton's time, coal for Lune Mills and St George's Works had been supplied by rail. However, towards the end of 1931, the new company was in negotiations with the LMS and the Commissioners with a view to bringing in their supplies by sea *via* Glasson Dock and storing some of it on the quayside. It seems that, by February 1932, the new arrangements were in place. The discharge of coal was accomplished utilising one steam grab crane owned by Williamson's and the two steam cranes with buckets, owned by the Port Commissioners and, of course, operated by the LMS. Unfortunately, the overwhelming majority of this coal was discharged into motor lorries to be taken to Lancaster. Speaking in March 1932, an LMS official commented that, whilst 9,000 tons of coal for Williamson's had recently been discharged into lorries at the port, only 200 tons had travelled up to Lancaster by rail. It is known that in the same year, Lancaster Rural District Council was obliged to spend £533 on reconstructing a culvert on the road between Glasson Dock and Conder Green. This had been in danger of collapse owing to the heavy lorry traffic carrying coal from the port to Lancaster. This change in the way Williamson's obtained their coal supplies must have represented a significant loss of traffic for the LMS. Most of the coal was shipped from Point of Ayr colliery near Prestatyn in North Wales, with much smaller quantities from ports on the Ayrshire coast in Scotland. Point of Ayr was situated next to the sea and the colliery had its own facilities for loading vessels. Satisfied with the new arrangements, at the end of March 1932, Williamson's applied to the Commissioners for permission to install two grab cranes, running on a dedicated seven-feet gauge line alongside the dock. In addition to this, the company wished to increase their coal storage area. In essence, the Commissioners and the LMS agreed to these proposals, with Williamson's meeting the cost of the alterations and the LMS carrying out the work to alter the existing sidings to accommodate the track for the cranes. The LMS also agreed to maintain the new track for the cranes alongside their responsibility for the existing sidings on the quayside. Details of what was actually provided do not appear to have survived, however, the Commissioners' minutes suggest that it closely followed Williamson's original proposals. The work appears to have been completed by around October 1932. Williamson's representative commented that the company had entered into a contract for the supply of 40,000 tons of coal by sea over a twelve-month period. It is a reminder that, following recovery from the dislocations and cost fluctuations brought about by the Great War, transport by sea could still provide a competitive alternative to rail transport.

Whilst the Port Commissioners had, in general, enjoyed at least constructive relations with the LNWR, the same cannot be said for its successor at Lancaster, the LMS. One of the major sources of friction was undoubtedly the maintenance of the two steam cranes at Glasson Dock. As we have seen, although these remained the property of the Port Commissioners, they were worked exclusively by the railway company and consequently the LMS maintained them at their own expense. This arrangement had been agreed with the LNWR back in 1883 and the railway company had always retained its own stock of spare parts for these machines. There is evidence to show that at the busiest times, around the turn of the century, the railway drafted in a third crane to assist with work on the quayside. However, by the late 1920s, this had long since disappeared to pastures new. The Commissioners' two remaining cranes were beginning to show their age, and the LMS seemed to be increasingly dilatory in keeping them both in working order. It is not difficult to see why. Whilst rail traffic to and from Glasson Dock remained light but steady, very little of this was generated by the dock itself. Glasson had become something of a quiet backwater, as fewer and fewer ships called to discharge their cargoes. For those that did, it was more often the case that the cargo was loaded into motor lorries rather than railway wagons. Matters came to a head towards the end of 1928, when one of the cranes broke down completely, seemingly requiring a new boiler. The absence of one of the cranes inevitably slowed things down when it came to discharging the limited number of vessels that still arrived at Glasson Dock. However, the LMS rather rubbed salt into the wound by charging the consignee demurrage on the wagons standing waiting on the quay side owing to the delay. This rather crass and clumsy response probably reflected more on the lack of initiative on the part of the LMS officials on the ground, than on the promulgation of official policy from on high. Nevertheless, it seems to have soured relations between the Commissioners and the shipowners on the one side and the railway on the other. In the event, the LMS provided a new boiler for the crane in question, but throughout the early 1930s there are sporadic references in the Commissioners' minutes to problems with the cranes at Glasson Dock, with the Commissioners insisting that both cranes be maintained in full working order. No doubt in the hope that one day there would be an increase in traffic and they would once again be put to full use.

Unfortunately, the relationship between the LMS and the Commissioners deteriorated further when the former claimed that, in fact, they owned a small portion of the quay surrounding the wet dock. This was the area surrounding the lock linking the dock with the canal basin. This particular dispute arose in 1929, when, with the dock generating very little rail traffic, the LMS began using the lines on the quayside to store empty wagons. This practice effectively blocked the usual berths on the east side of the dock and so any visiting vessel was obliged to discharge its cargo in the corner close to the entrance to the canal basin. Here, so the LMS claimed, the cargo was placed onto land owned by them and, therefore, additional wharfage

This is one of an earlier batch of 14-ton tank wagons constructed in 1930 for Williamson's, again by Charles Roberts. The number of wagons ordered is not recorded but the number 21 suggests a fleet of some size. Clearly the wagon is designed for the carriage of inflammable liquid and in the case of Williamson's this is likely to be the white spirit used in their manufacturing processes. HMRS.

charges were due. To be scrupulously fair, the issue of the boundaries between the land owned by the Commissioners and that owned by the canal and therefore by the railway company, was a little confused. Back in LNWR days, there had been some doubt about this, but after amicable discussion on the subject, an agreement had been reached, and boundary markers had been put in place. In an attempt to gather evidence to resolve this current dispute, both organisations delved into their archives. The Port Commissioners, established by an Act of Parliament in 1749 and thus predating even the era of canal construction, perhaps had the advantage here. The Commissioners were convinced that they had discovered a document that confirmed their ownership of the relevant portion of the quayside. However, the LMS disagreed! The dispute dragged on in a desultory fashion until April 1935, when the boiler in the other steam crane at Glasson Dock finally succumbed to the ravages of time. The LMS, rather cheekily, wrote to the Commissioners, asking them to provide and to install a new boiler in the crane. This was, of course, contrary to the agreement between the Commissioners and the LNWR dating from 1878. The Commissioners' solicitor, Mr Satterthwaite, wrote to the LMS pointing this out and holding them responsible for any demurrage arising as a result of the crane being out of action. However, the LMS would not repair the crane. Perhaps spurred on by the railway company's intransigence, the Commissioners, or rather their solicitor, took a closer look at various Acts of Parliament and agreements which, over the years, had had some bearing on the relationship between the canal, the railway company and the Commissioners. In August 1936, they concluded that over the years, certain annual payments should have been made by the owners of the canal (i.e. the LNWR and its successor the LMS) to the Commissioners, and that these were outstanding. This was supported by Counsel's opinion obtained in 1864, at the time the LNWR took out their original lease on the Lancaster Canal. The Commissioners, therefore, were able to present a bill to the LMS for £1,310. 19s.1d. representing the accumulated arrears of these payments.

This further escalation of events seems to have brought both sides to the negotiating table. Towards the end of 1937, some sort of agreement had been reached. In essence, the LMS would insert clauses in their forthcoming omnibus bill for the 1938 parliamentary session, which would extinguish the responsibilities imposed upon the railway company in relation to the operation of Glasson Dock. These, of course, originated in the agreement of 1878 and in the Act of the same year authorising the construction of the railway. The wording of the relevant clauses in the bill was agreed with the Commissioners who, for their part, undertook not to oppose its passage through Parliament. The question of the boundaries at Glasson Dock appears to have taken rather longer to resolve. By the Summer of 1938, some sort of tacit understanding had been reached. However, it appears that the resulting agreement was still awaiting a final signature as late as September 1939. The London Midland and Scottish Railway Act of 1938 received royal assent on the 2nd of June 1938. Shortly after this, the LMS wrote to the Commissioners, proposing that they would relinquish all responsibility for the operation of the dock and the maintenance of the sidings and equipment with effect from 1st August. This was agreed to by the Commissioners. Goods trains continued to run through to Glasson when required, of course, and these trips would include occasional forays onto the sidings alongside the wet dock and the river quay. However, the ongoing maintenance of these was now solely in the hands of the Commissioners. Williamson's were still bringing most of their coal into Glasson Dock, of course, but this was leaving the port by road. It is likely that this was also the case for the dwindling number of other vessels still discharging their cargoes at the dock. As a parting shot in the whole acrimonious affair, the LMS presented the Commissioners with a bill for £18, representing the additional cost of maintaining the extra set of rails used by Williamson's cranes. This was promptly repudiated by the Commissioners and the railway company was obliged to apply to Williamson's themselves for payment. It also appears that the old steam crane never did receive its new boiler. However, its twin, still apparently in working order, must have sat idle on the quayside for significant periods of time. Before the break came, the LMS were being asked to maintain and to pay rental in respect of appliances and sidings that did not belong to them and from which they derived little or no income. Perhaps, ultimately, the Commissioners, most of whom were also local businessmen, saw the financial wisdom of the railway company's repudiation of the 1878 agreement and acquiesced in the new arrangement. With the LMS no longer involved, there was now no one available to load or unload any vessels that came to Glasson Dock. Ship owners or consignees would, therefore, have to make their own local arrangements. This can only have acted as a further discouragement for anyone wishing to use the port. By August 1938, shipowner Robert Gardner (and incidentally, now a Port Commissioner) was the only regular user of the facilities at Glasson Dock. He immediately came to an arrangement with the Commissioners to operate the surviving crane himself, but entirely at his own risk. With the LMS now freed from responsibility for the arrangements at the wet dock, the question arose as to how much they should be charged for using the lines on the quayside. There are no details of any figures in the Commissioners' minutes, although by this time this is likely to have been an infrequent occurrence. Nevertheless, it did happen from time to time and in November 1938, the Commissioners wrote to the LMS pointing out that they had been using the lines along the quayside without permission.

Although Glasson Dock was now distinctly quiet, at the Lancaster end of the branch, business seems to have been fairly brisk and traffic remained at satisfactory levels. The gas works still required regular supplies of coal, all brought in by rail, of course, but the largest source of traffic by far was Williamson's. After the early 1930s, which was a difficult time for much of British industry, the company settled down to a period of relative prosperity. Linoleum remained a popular product both at home and abroad. It was considered modern and hygienic and it was the floor covering of choice for the increasing number of bathrooms and indoor WCs, especially those provided in the myriad of modern semi-detached houses that were so much a feature of the inter-war housing boom. By 1938, British exports of hard flooring had reached 80% of their pre-1914 level. This was a significant achievement and it does appear that Williamson's played their part in this. Although most of the raw materials were now coming into Williamson's works by sea or by road, the expanding range of finished products still left Lancaster by rail. One feature of the outgoing rail traffic from Lune Mills was the biannual issue of Williamson's book of trade patterns sent to agents in the UK, as well as throughout the Empire and much of Europe. Each issue necessitated the provision of a dedicated train of parcels vans from Lancaster. The event was reported in the local newspapers and the LMS, always quick to see a promotional opportunity, provided the locomotive with a headboard bearing the legend *Lancaster Linoleum Pattern Special*. A photograph of one of these trains featured in the *Lancaster Guardian* for 3rd September 1937.

It is known that, in the early 1930s, Williamson's purchased a number of modern 14-ton capacity tank wagons from the Yorkshire railway wagon builder, Charles Roberts & Co. Photographic evidence confirms that the first batch, supplied in 1930, carried the mandatory livery and lettering for tank wagons conveying inflammable liquid. For Williamson's this is likely to have been the white spirit used in their manufacturing processes. It is not known how many of these wagons were constructed, however, the highest running number recorded was 21, suggesting a fleet of significant size. Prior to the construction of these wagons, the white spirit seems to have been delivered in barrels, with at least some of it arriving by sea. A year later, in 1931, a further batch of six tank wagons was constructed by Charles Roberts. These were of a slightly different design and carried a different livery, boldly advertising the company's cork linoleum. Given the change of livery, these were clearly not intended for the conveyance of inflammable liquid and the most likely alternative would be the linseed oil used in large quantities in the manufacture of linoleum. However, around this time, Williamson's were also bringing this commodity in by sea and discharging it at the New Quay. Indeed, in 1933, the company obtained permission from the Port Commissioners to install a pipe on the New Quay so that linseed oil could be pumped from the holds of ships, straight into storage tanks on the Lune Mills site. Nevertheless, there is nothing to preclude the use of both modes of transport if this provided the best business outcome for the company.

Chapter Eight

War Again and the Decline Towards Closure

BRITAIN DECLARED WAR on Nazi Germany on 3rd September 1939. As in the Great War, the railways came immediately under government control and were run by the Railway Executive Committee, which included managers from each of the four large companies. Unlike in 1914, it had been clear for some time that war was imminent. In fact, the relevant order establishing government control was brought into force by the Minister of Transport on 1st September, 48 hours before the official declaration of hostilities. Owing mainly to the effectiveness of wartime censorship, it can sometimes be quite difficult to ascertain exactly what happened on the railway at a particular location during the conflict. The Glasson Dock branch is no exception to this particular frustration. However, by drawing on a range of sources, it is possible to assemble at least a partial picture of what went on. Certainly, the Admiralty seems to have had some sort of presence at Glasson Dock from very early on in the war. Williamson's coal unloading and storage operations were somewhat curtailed towards the end of 1939 and instead, arrangements were made for the company to store coal on the New Quay at Lancaster. By 1942, a portion of the quayside at Glasson Dock had been fenced off, seemingly with wire fencing topped with barbed wire. Access into the enclosed area was controlled by security police. It is known that several vessels from the Dutch navy, including the destroyer *Van Galen*, spent some time at Glasson Dock before being sent to the Far East. Local memory recalls that on occasions, both the wet dock and the canal basin were packed with Royal Navy high-speed motor launches, which were used for air-sea rescue work in the Irish Sea. In December 1944, HMS MFV 138 (a naval motor fishing vessel), collided with the wooden jetty at the entrance to the wet dock, causing £200 worth of damage. The bill for the repairs was met by the Admiralty. All of these snippets of information suggest a sustained naval presence at Glasson Dock throughout the war. However, the dock was not entirely closed to normal commercial traffic and some cargoes were still discharged at the port, the vessels tending to be the larger ones, which would have had difficulty negotiating the river channel up to Lancaster.

From around February 1941, timber importers May & Hassell Ltd began to store timber at Glasson Dock and, by the Spring of that year, they were renting 3,600 square yards of space from the Commissioners. The company was a large one, with branches at several ports, the nearest one to Glasson Dock being Liverpool. It is perhaps significant that this time frame coincides with the Liverpool blitz that culminated in the first week of May 1941 with a sustained bombardment over several days, which devastated the city. It may be that the company made the decision to move their operations further away from the attentions of enemy bombers and the quiet backwater of Glasson Dock fitted the bill perfectly. Certainly, in January 1941, May and Hassell were in the process of opening a branch office in Lancaster and were advertising in the local newspapers for a timber clerk. Local memory has it that the quaysides at Glasson Dock were piled high with stored timber during the war and this is obviously what was recalled. By July 1942, the amount of storage space rented from the Commissioners had reduced to 360 square yards. However, in June 1943, the company entered into an agreement with the LMS for the tenancy of an area at Glasson Dock. May & Hassell continued to store timber at the dock throughout the war years and it is known that at least some of it travelled either in or out by rail.

In January 1941, the Commissioners contacted the LMS to ask what it would cost to repair the sidings at Glasson Dock. The price quoted by the railway company was £75. It is doubtful that these repairs were carried out as, in December of the same year, the LMS wrote to the Commissioners complaining that one of their wagons had been derailed on the quay lines, resulting in damage amounting to £7. 7s.6d. The Commissioners' response was to point out that the railway had been using the lines on the quayside without permission and to refuse to pay for the damage. Arguably the LMS were taking a chance using the poorly maintained sidings at Glasson Dock with or without permission. It is perhaps fortunate that wagons were always propelled onto these lines and, therefore, the much heavier locomotives did not have to venture too far onto them. All of this suggests that the lines on the quayside had received little or no attention since the LMS relinquished responsibility for them in 1938.

A view looking towards Lancaster with Lune Mills on the left and Williamson's power station on the right. The latter was commissioned in 1949. This was taken from the occupational crossing and the two separate cattle creeps situated where the Freeman's Wood footpath crossed the line. The buffer stop denoting the end of the quay goods line is visible in the distance.

CRA, Rev J Jackson Collection.

A busy scene at the rear of Lune Mills taken around 1945. On the right is the rear wall of the site with the two parallel tracks of the route to Glasson Dock (to the left) and the Lancaster Quay goods line (to the right). The area just to the left is taken up by the War Department sidings put in at the beginning of 1943. A portion of this area was covered by Williamson's new power station, which was constructed between 1946 and 1949. Courtesy Lancaster Museums.

If the LMS had washed their hands of the Commissioners' lines on the quayside at Glasson Dock, the same could not be said for their own facilities at the Lancaster end of the branch. Here, on the two lines running at the rear of Lune Mills, the railway company provided a number of additional features. In March 1941, a second crossover was brought into use, connecting the main running line to the quay goods line. This was situated roughly halfway between the junction for the Ford Quay and the Freeman's Wood end of the long head shunt. This, for the very first time, provided run-around facilities for traffic moving to or from Lune Mills and the Ford Quay. At the beginning of 1943, the LMS constructed a group of four sidings on the vacant marsh behind Lune Mills. These appear to have been put in at the behest of the War Office. Rail access was from the main running line *via* a single trailing connection for trains travelling towards Glasson Dock. It seems that the original intention was to have three roads only, as an LMS plan dated 17th February 1943 titled *Proposed Additional Siding and Crossover Road at Messrs Williamson & Son's Siding*, shows an additional set of rails added to the existing group, making four sidings in all. There is a slight mystery concerning these sidings insofar as, in at least one LMS document, they are referred to collectively as the ASR Depot. What ASR stands for is unclear, although it is perhaps tempting to link the sidings with the activities of the Royal Navy's air-sea rescue launches that were stationed at Glasson Dock. However, it has not proved possible to uncover any evidence to connect the two. The LMS plan also shows the provision of a third crossover from the main running line to the long siding at the rear of Lune Mills. This was situated towards the Freeman's Wood end. This addition transformed the siding into a rather long loop with an intermediate crossover. This must have made the shunting of traffic on the Lune Mills and the Ford Quay lines much easier. The title of the LMS plan suggests that, from the outset, the new fan of sidings was used for traffic to and from Williamson's. Photographs taken around 1945 reveal that much of the area was covered by a large, open sided shed, effectively a sort of roof on stilts, which enabled wagons to be loaded or unloaded under cover. A trawl through the local newspapers suggests that, from around 1943 at the latest, some sort of war work was being carried out at Lune Mills and that this employed significant numbers of female workers. Unfortunately, owing to the customary efficiency of wartime censorship, it has not proved possible to discover exactly what the nature of that work was. However, it does appear that these covered sidings were used for the reception of the necessary materials and for the dispatch of outgoing traffic.

Also appearing on railway plans for the first time during this period, is a short single siding alongside the eastern end of the rear of Lune Mills, with a trailing connection for trains travelling towards Glasson Dock. The precise date this was brought into use is not known. As we have seen, in 1930–31, Williamson's acquired a number of tank wagons and this may provide a clue, although the siding does not feature on the 25-inch Ordnance Survey revised to 1931. Certainly, photographic evidence and local memory indicate that the siding was used for the reception of railway tank wagons carrying tar or white spirit. The tar was used as a coating on the heavy backing paper, which, at this time, was utilised in the cheapest grades of linoleum. Both commodities were pumped or siphoned into storage tanks within the works, the tar being reheated by means of a steam lance. Access to the siding was *via* a door at the rear of the site and portable ladders were used to access the pumps and pipework on the boundary wall.

The Second World War finally ended on 15th August 1945 with the surrender of Japan. As things began to slowly move back towards some semblance of normality, Williamson's embarked upon a significant programme of modernisation of the plant and facilities at Lune Mills and this initiative was to have an impact on the railway infrastructure serving the site. As early as May 1946, the board of directors had authorised the construction of a railway siding running into the packing and dispatch department warehouse. This was planned in conjunction with the LMS and was brought into use on 24th November 1947. Rail access into the site was at the north east end, from the single line running onto the Ford Quay. Documents held in the Williamson's deposit at Lancashire Archives indicate that there were two sidings within the warehouse and that these were cleared of loaded vans twice daily. By arrangement, the manager of the warehouse department would notify the railway, in advance, of the number of vans required for loading each day. It is known that on the first day of operation of the new siding, a total of 37 vans of linoleum-related products were dispatched from Lune Mills over two shunts.

Fowler 4F No 4032 stands at the remains of the platform at Glasson Dock station around the time of nationalisation. Other images taken at the same time show the train to be composed of two vans and two brake vans, one of which was a six-wheeled ex-Caledonian Railway vehicle.

CRA, Rev J Jackson Collection.

However, by far the largest of Williamson's post-war projects was the construction of a dedicated power station supplying electricity and high-pressure steam to the whole of the Lune Mills site. This was situated at the rear of the site on the opposite side of the branch line. Prior to this, steam for various purposes had been supplied to Lune Mills *via* a large number of individual Lancashire boilers scattered across the site. The existing electricity supply came principally from Lancaster power station, but was supplemented by an onsite, steam-driven, direct-current powerhouse dating from the early years of the twentieth century. Work commenced in 1946, and the final cost was just a little short of one million pounds. The facility commenced operation in September 1949 and was expected, ultimately, to reduce the company's coal consumption by around 20,000 tons *per annum*. The ex-War Department sidings were utilised, first of all for the delivery of materials for construction and then later to supply coal to the site. It was necessary to alter the layout and reduce the number of sidings in order to accommodate the footprint of the new building. Aerial views taken shortly after the opening, show a single siding running alongside the power station with two shorter sidings terminating at the western end of the building. The commission of the power station also brought about a complete change in the way Williamson's obtained their coal supplies. These were now brought in by rail, and the longest siding served a coal tippler and skip conveyor, which emptied each wagon and transferred the coal directly into the power station's bunkers. It is not known where Williamson's sourced their coal supplies in the 1940s and 50s. However, by 1961, they were coming

40/15/20

Serial No. 1

JAS. WILLIAMSON & SON, LTD., LUNE MILLS, LANCASTER.

DAILY RAIL VAN REQUISITION.

Requisition for the First Train for Despatching Goods from LUNE MILLS.

Date Ordered	Date Required		Number Required No. 1 Siding	No. 2 Siding
21st Nov/47	24th Nov/47.	1st Shunt	25	
		2nd Shunt	12	
		Total	37	

[signature] Manager, Warehouse Dept.

[signature] PRODUCTION DIRECTOR.

Internal paperwork from Williamson's that was used on a daily basis to requisition the required number of railway vans for outgoing traffic. This represents the first shipment from the newly constructed siding into the dispatch warehouse and consequently has been preserved by the company.

Courtesy Lancashire Archives.

A casual observer might be forgiven for thinking that the passenger service to Glasson Dock has been restored. In reality this is an enthusiasts' special, *The North Lancashire Rail Tour*, which visited several branch lines in the area on 1st May 1954. The photograph captures one of a series of shunting moves by which the train was divided so that the Fowler tank could run around the whole in preparation for the return journey. CRA, Pearsall Collection.

from Lofthouse Colliery near Wakefield. BR working timetables from the early 1960s show a regular twice-weekly working, from Stourton (Leeds), to Lancaster Castle. A photograph, taken in June 1961, shows one of these trains composed of fitted BR 16-ton mineral wagons and in charge of Ivatt-designed 4MT 2-6-0 No 43052. The wagons themselves appear to have been branded to run between Lofthouse Colliery and Lancaster only. The train is facing south and standing in the throat of Castle Yard, just to the south of the station. In due course the train will reverse direction and run briefly northwards through the station before taking the Glasson Dock line and commencing the steep curving descent down to Williamson's power station.

The run of Commissioners' shipping registers held at Lancashire Archives ends in 1946, so from this point on, there is no longer a clear picture of the amount of traffic using the port as a whole. Glasson Dock itself seems to have been especially quiet after the war, although some sort of naval presence appears to have been retained for a short period after the cessation of hostilities. In the Spring of 1946, there was a limited resumption of timber traffic into the dock, the recipient being May & Hassell, who had been storing timber in the quayside since the early years of the war. This was to be taken from the port by rail and, in June of that year, the LMS wrote to the Commissioners requesting permission to use their sidings for this purpose. Permission was readily granted. However, the Commissioners wrote to both the LMS and to May & Hassell warning them that this was on the understanding that they used the lines entirely at their own risk and responsibility. This perhaps speaks volumes about the condition of the track on the quayside by the end of the war. Some remedial work must have been carried out shortly after this as, in November 1946, the Commissioners noted that the LMS was now prepared to shunt down the lines serving the river quay at Glasson Dock. This being the case, they decided to offer the use of a ship-breaking berth to T W Ward of Sheffield and also to Nicholson's who operated the local shipyard and graving dock. However, there appears to have been little interest from these two concerns and in the event, it was agreed that the berth would be let to Messrs T Grubb for £200 *per annum* for three years. Unfortunately, the ship-breaking venture does not appear to have taken off and Messrs Grubb subsequently withdrew from the agreement. However, they did moor at least one vessel at the berth, for which they were charged the usual dock dues.

In the July 1945 General Election, the Labour Party had been swept to power in a landslide victory, and fairly quickly, the new government implemented a series of schemes for the nationalisation of strategically important industries. The railways were amongst the earliest to be considered and, in November 1946, the Government submitted a transport bill to Parliament, which became the Transport Act of 1947, receiving royal assent on 6th August of that year. The Act created the British Transport Commission (BTC), which was intended to create an efficient and integrated transport system across the whole of the UK. Each of the different transport sectors was to be managed by its own executive branch of the BTC. The management, operation and maintenance of the railways themselves was delegated to the Railway Executive. The railways passed into public ownership on 1st January 1948, and the Glasson Dock branch became a part of British Railways' London Midland Region, the title being the business trading name of the Railway Executive of the British Transport Commission.

The London Midland Region does seem to have been very quick off the mark in examining the financial viability of the less well used portions of their new empire. The first official line closure under British Railways took place as early February 1948, and the victim was the goods-only line from Mantle Lane East to the foot of the Swannington Incline in Leicestershire. Then, in May of that year, the first passenger service withdrawal took place between Woodford and Hinton on the Great Central route and Byfield on the Stratford and Midland Junction Railway. Both of these closures took place under the auspices of the London Midland Region. It wasn't long before someone decided to have a closer look at the Glasson Dock Branch. In June 1948, the Port Commissioners were approached by Mr JD White, the BR Traffic Manager at Barrow, to ascertain what their view would be in the event of the railway seeking to close the line between Lancaster and Glasson Dock and lifting the lines for use elsewhere.

Another view of Glasson on the occasion of the visit by *The North Lancashire Rail Tour* on the 1st May 1954. The photographer has climbed up one of the signals protecting the level crossing to obtain this slightly unusual perspective on the station area. The level crossing gates appear to have been freshly painted. The area to the left was the site of the station's tiny goods yard. There had been a small platform here with an end loading facility, but this had been removed by this date. L&NWRS.

Discussing the matter amongst themselves, the Commissioners decided that the suggestion could not be entertained. In any case, there was a distinct possibility that a number of cargoes of timber might be dispatched from the dock by rail in the near future. There was, however, a further possible complication. The original 1878 agreement incorporated in the Act of Parliament for that year had imposed an obligation on the railway to maintain the line to Glasson Dock, effectively in perpetuity. Therefore, a new Act would be required to repeal the relevant provisions of the original. Only then could the line be closed. Whilst this would certainly not present an insurmountable difficulty for the Railway Executive, a repeal of the 1878 Act might well create an additional problem for the Commissioners. By virtue of certain clauses in it, and in the 1894 Furness Railway Act, the Port Commissioners, for many years, had been entitled to a share of the shipping dues arising from the lighthouse they maintained on Walney island, close to the port of Barrow-in-Furness. This amounted to a significant source of income that was used to maintain the port facilities at Glasson Dock and Lancaster. The Commissioners were concerned that if the 1878 Act was repealed or altered in some way, this might affect their entitlement to these dues. With all of this in mind, Mr White was informed that should the railway decide to close the line, the Commissioners would oppose this with all the means in their power. This does seem to have put an end to the threat of closure, at least for the time being. Perhaps the decision-makers at BR were influenced by the developments taking place at Lune Mills, which promised a significant increase in traffic at the Lancaster end of the branch.

Notwithstanding the Commissioners' stout defence of the line, activity at Glasson Dock itself appears to have remained at a low level. In December 1948, it was reported that work had begun to remove a line of railway along the north easterly side of the wet dock. This appears to have been the third siding put in by the LNWR in 1903. By March 1949, this had been completed. The redundant rails were to be cut up and put to various uses around the dock estate. However, there remained about 400 cast iron chairs, together with fishplates and screws, and these were sold to a local scrap dealer.

During the 1950s, there were occasional references in the local newspapers to cargoes arriving at Glasson Dock. In August 1953, a large consignment of bricks arrived from the Isle of Man. However, this was a response to a particular shortage in the Lancaster area. Then, in June 1954, it was reported that a cargo of 500 tons of linseed oil, presumably destined for Williamson's had been discharged at the dock. That such events were reported in the newspapers suggests that they occurred somewhat infrequently. Furthermore, there is no evidence that either of these commodities left the port by rail. Local memory has it that during the 1950s, a special train was run from the Wigan area to Glasson Dock, comprised of around 30 wagons loaded with what were described as oil drums. The details are slightly vague, but these are said to have contained an unspecified substance destined for the export market. The engine, a member of the ex-LMS Patriot class, was crewed by men from Wigan who did not know the road on the Glasson Dock branch. Consequently, a driver and guard from Lancaster Green Ayre acted as pilots for the final leg of the journey. Owing to the length of the train and the presence of a large main line locomotive, some difficulty was experienced in running around the wagons at the terminus, so that they could be pushed onto the quay lines. As a further sign that there was still some railway traffic generated by the dock, in February 1953, the Harbour Master reported that he had arranged to have the lines to the pier head cleared of obstructions and debris. This was to allow a cargo of basic slag to be discharged into a total of 72 wagons. Given the number of wagons involved and the limitations of the railway facilities at Glasson Dock, it is likely that at least two return trips down the branch were necessary to clear the cargo.

In December 1954, the sawmill at Glasson Dock, still operated by the Westmorland Cooperage and Sawmills Co, finally closed. During the 1920s this had been the largest source of railway traffic at the Glasson end of the branch. However, by the time closure came, the firm had curtailed its operations significantly and appears to have been using road transport rather than rail.

On 24th June 1955, the Lancaster Guardian reported on a possible revival of trade at the port, citing the loading of three coasting vessels with nitrochalk (calcium ammonium nitrate) over

A slightly unusual view of Glasson station building taken from the road into the village. Access onto the platform was *via* a wicket gate In the fence. With a carriage standing at the platform, it is not difficult to imagine the station as it was in its heyday with a short rake of LNWR six-wheeled carriages waiting for the next train to Lancaster. Taken on 1st May 1954.

L&NWRS.

the last few weeks. This was a form of fertiliser manufactured by Trimpell Ltd at their plant in nearby Heysham. However, this traffic does not appear to have continued. Furthermore, there is no evidence that the nitrochalk arrived at the port by rail, although this is a distinct possibility. Similarly, during this period, Trimpell Ltd made use of the port to send out shipments of coke breeze dust. However, this appears to have been brought from Heysham by road and stored on the quayside.

In November 1957, the *Morecambe Guardian* ran a story (which included a rather indistinct photograph) concerning a short passenger train that had arrived at Glasson Dock, carrying workers who were engaged in laying a section of new track. Perhaps it was a quiet week for news in the offices of the *Morecambe Guardian*. Local memory recalls that the engine in question was Ivatt 2MT 2-6-0 No 46441. The newspaper commented that an occasional freight train came to Glasson Dock when required, but there had been no passenger service for 17 years, with the exception of one run two years earlier for the benefit of a railway society. In fact, the last passenger train had been on the 1st May 1954, when the line was visited by a tour organised jointly by the Stephenson Locomotive Society and the Manchester Locomotive Society. Titled *The North Lancashire Rail Tour*, this was hauled by ex-LMS Fowler 4P 2-6-4 tank No 42316 and visited several branch lines in the area. At Glasson Dock, the seven carriages had to be split into two sections to enable the locomotive to run around the train in preparation for the return journey. There was a further visit by a passenger train on the 29th May 1960, when Fairburn 4P 2-6-4 tank No 42136, together with Stanier 6P5F 2-6-0 No 42952, were seen at Glasson Dock on *The Northern Fells Rail Tour.* This was another event organised jointly by the Stephenson and the Manchester Locomotive Societies and the itinerary covered a wide range of routes stretching from Ulverston in the west to Kirkby Stephen in the east and travelling as far north as Penrith.

By 1960, the railway facilities at Glasson Dock were looking distinctly shabby. On much of the branch, the permanent way appears to have been relatively well maintained and kept reasonably free from weeds and encroaching vegetation, especially at the Lancaster end. However, as Glasson Dock was approached, the weeds and grass became more prevalent and during the Summer months, the rails in the station area seem to have all but disappeared under a carpet of greenery. Photographs taken at different dates during the 1950s suggest that periodically work was carried out to remove at least some of this weed infestation, and BR papers indicate that the permanent way between Lune Mills and Glasson Dock was re-laid around this time. At Glasson, the timber station building facing the estuary and subject to the full force of gales coming in from the Irish Sea began to look increasingly dilapidated. At some point, the small adjacent building, which seems to have housed a lamp room and stores, was demolished and most of the passenger platform was removed, leaving only the section directly in front of the station building. Photographic evidence suggests that these alterations took place prior to nationalisation.

Another view of the station on the occasion of the visit by *The North Lancashire Rail Tour*. The photographer is looking towards Lancaster and the station building is visible beyond the level crossing gates. Given the limited facilities at Glasson, it proved necessary to divide the seven carriages into two sections and place one of them temporarily on the line to the river quay. This enabled the locomotive to run around the train in readiness for the return to Lancaster

L&NWRS.

In the early 1960s, Ivatt 2MT No 46422 was a regular performer on the branch. Here she is shunting vans into the warehouse at Lune Mills, while the shunter looks on. The quay branch continues on across New Quay Road and onto the sidings serving the river quay itself and the gas works. Taken on 11th March 1964.

Ron Herbert.

On the Lancaster Quay goods line, the gasworks ceased production of town gas, seemingly in 1959, following the opening of the new gasworks at White Lund (between Morecambe and Lancaster). However, the site on the Ford Quay was retained and the existing gas holders continued in use for storage purposes. Perhaps for this reason, the private siding and its connection with the lines on the quay remained *in situ* for the time being. Up until the end of gas production on the old site, there were regular trains of up to a dozen wagons of coal. These were placed in the private siding by a BR locomotive. However, local memory recalls that latterly, the gasworks used a mechanical tractor to pull the wagons into the premises. It is also recalled that from time to time in the 1950s, there was some outgoing traffic in railway tank wagons and is likely to have been for tar or ammoniacal liquor. Another link with the past was broken in 1960, when Messrs W & J Pye relinquished their lease on the shed on the Ford Quay, giving six months' notice effective from 12th November. However, it is doubtful that the firm was making any use of the rail facilities by this time. Notwithstanding this reduction in the level of commercial activity, the quay goods line remained relatively busy owing to the activities of Williamson's. Several sources recall long trains of vans being taken from the warehouse at Lune Mills and up the bank to Castle Station in the 1950s and early 60s.

During the period 1962–63, the sidings on the Ford Quay were the scene of significant activity, when they were utilised to supply materials for the reconstruction of the Carlisle railway bridge, which carried the main line over the River Lune. The spans across the river itself were rebuilt using composite steel and concrete construction. The new bridge was formed of four main longitudinal steel girders, and utilised cantilevered steel girders to carry the concrete deck slab. In April 1962, the Commissioners came to an agreement with the contractors, the Cleveland Bridge & Engineering Co Ltd, for the use of a portion of St George's Quay at a rental of £5 per week. The large girders were brought down the branch on bogie bolster wagons and placed in the two sidings on the quay. The contractors constructed a purpose-built lifting facility, which spanned the sidings. This enabled the girders to be removed from the wagons, raised to the height of the bridge and then rotated so that they were parallel to it.

In the mid-1950s, as we have already seen, there had been a limited number of cargoes of nitrochalk going out from Glasson Dock. These shipments seem to have re-occurred sporadically, as BR correspondence dating from June 1961 refers to what was probably a single shipment of 671 tons of this material being taken to the port by rail. However, in 1962 there began what seems to have been a reasonably regular series of shipments, all of which appear to have been brought to the port by rail. On the 2nd March 1962, Ivatt 2MT 2-6-0 No 46422 was noted at Glasson Dock, having arrived from Heysham Moss sidings with a special train of nitrochalk destined for the Irish Republic. After some shunting and remarshalling of the train in the station, it was split into two and both rakes of wagons were propelled into the sidings alongside the wet dock. Towards the end of 1962, the Commissioners had put their own men to work clearing the lines alongside the wet dock so that they could be fully utilised for this traffic. By February of the following year, the Harbour Master could report that the railway lines to the pier head were now completely open. Shunting had taken place all the way along the dockside and this had enabled the loading of vessels to take place with the least possible delay. It is known that on the 10th September 1963, a similar train ran from Heysham Moss to Glasson Dock, again carrying nitrochalk destined for export to Ireland. On this occasion, the locomotive in charge was No 42063, a Fairburn-designed 4P 2-6-4 tank engine. Such was the level of traffic that ICI, who were now operating the plant at Heysham, brought in a mobile conveyor to assist with the loading of vessels. Local memory suggests that during this period, there were other special trains of this nature, which went unrecorded. It is known that during the year ended 31st December 1962, a total of 6,390 tons of nitrochalk was exported from Glasson Dock. If all of this arrived at the port by rail, which seems highly likely, the figure represents a significant number of special trains within the twelve-month period.

In 1963, a ship-breaking company with the slightly unusual name of Lacmots Ltd, established itself on the river quay at Glasson Dock, taking over two of the three available berths and utilising the two railway sidings there. This concern already had a similar, larger facility at the port of Queenborough, on the Isle

An excellent view of the railway facility serving the ship-breaking berths at Glasson Dock that, from March 1963, were operated by Lacmots Ltd. The photographer is standing on the veranda of the brake van at the rear of the train that has just deposited the withdrawn Fowler tank No 42301 in the sidings.

Ron Herbert.

of Sheppey, in Kent. By March 1963, the first two vessels had arrived for breaking up and work seems to have commenced in the following month. By August of that year, the company was using its own steam crane on the lines alongside the river quay. However, it was not until October of that year that the fine detail of a longer-term lease was agreed. Lacmots agreed to install, at their own expense, a point to the lines on the quay and that this would be left *in situ*, free of cost at the expiry of the lease. In December 1963, it was reported that the firm had purchased and laid about 60 feet of track on the river quay and had also replaced some rail that was badly worn. By the end of the year, about 300 tons of scrap had been dispatched from Glasson Dock by rail and almost 100 tons by road. However, in March 1964, the company seem to have switched to shipping for the removal of scrap from the site. On 14th March, the MV *Novel* sailed for Rotterdam with 600 tons of scrap. This was noteworthy as it was the first export to the continent from Glasson for many years. There were similar shipments of 400 and 600 tons to Rotterdam in June and July of that year. Lacmots Ltd was also in the business of scrapping redundant railway locomotives and rolling stock and it is known that in March 1964, a Fowler 2-6-4 tank engine No 42301 was taken to Glasson Dock for cutting up.

The movement was carried out as part of a normal trip working on the branch and the dead locomotive was marshalled next to the train engine, which was, once again, Ivatt 2MT No 46422. This was a Green Ayre engine and does seem to have been a regular performer on the branch during this period. The Fowler tank was left on the Ford Quay while 46422 carried out some shunting at the private siding serving the dispatch warehouse at Lune Mills. The locomotive's final journey to Glasson Dock was made in the company of a rake of fitted vans from Williamson's. At Glasson, 46422 ran around its train and then propelled it into Lacmots' sidings with the dead locomotive foremost.

Notwithstanding this increase in rail traffic to and from Glasson Dock, in May 1963, the Commissioners received a letter from the British Railways Board enquiring whether there was any prospect of traffic to and from Glasson Dock being developed in the foreseeable future. If not, the letter asked, what would the Commissioners' attitude be to the closure of the railway line, bearing in mind that the facilities were provided under the terms of the agreement with the LNWR dated 21st June 1878 and scheduled to the LNWR Act of that year. The Commissioners replied that they felt that it was essential that Glasson Dock be served by both road and rail and consequently they would resist any attempt to close the branch line. This was a tactic that

It is March 1962 and Ivatt 2-6-0 46422 crosses the Conder viaduct as it makes its way back towards Lancaster with a train of empty open wagons. The tide is out, revealing a vast expanse of salt marsh.

Ron Herbert.

A train of fitted vans is propelled slowly across New Quay Road towards Lune Mills. In the early 1960s, Williamson's and, then from 1963, Nairn-Williamson Ltd were regularly sending out a significant number of vans each working day loaded with rolls of linoleum or vinyl floorcovering. In addition, there were twice-weekly inbound workings of fitted mineral wagons carrying coal for their power station.

Ron Herbert.

had proved effective in 1948. However, since then, the political landscape had changed somewhat. The 1962 Transport Act had broken up the British Transport Commission and established the British Railways Board (BRB). The new Board was directed that the railways should be run in such a way that their profits should be "not less than sufficient" to meet their operating costs. What this meant in practice was that each service should pay for itself or at least be able to demonstrate the potential to be able to do so. The days of profitable routes supporting the unprofitable ones with the taxpayer making good any discrepancy were over. At least that was the theory. The Act also created a streamlined process for the closure of railway lines, which would be put to good use in the years to come. One of the products of these changes, of course, was the famous Beeching report which was published in March 1963. Therefore, it was against this background that the Commissioners received their initial enquiry from BRB in May 1963.

We have already seen that the Commissioners had been concerned that a repeal of the 1878 Act and the closure of the branch, in some way, might effect the port's entitlement to a share of the shipping dues arising from their lighthouse on Walney Island. However, after going into the matter thoroughly and seeking advice, they were reassured that this would not be the case. Nevertheless, in the Spring of 1963, the line to Glasson dock was still being used fairly regularly for the nitrochalk shipments and the Commissioners, quite reasonably, feared that closure would put an end to this traffic. Early in September of that year, the Commissioners received a letter from Mr Stirk, BR's District Goods Manager at Bolton, requesting a preliminary interview to discuss the implications of the railway's desire to close the line to Glasson Dock. In the previous month, 1,010 tons of nitrochalk had been carried on the railway to be loaded into vessels at the port. This, coupled with the railway's ongoing responsibility to maintain the line under the 1878 LNWR Act, gave the Commissioners at least some hope of staving off closure of the line. Perhaps with a view to mobilising some support against the depredations of the railway authorities, it was decided to report the proposed closure to the Ministry of Transport, the Dock and Harbour Authorities Association and ICI at Heysham, who, of course, were using the line for their shipments of nitrochalk. On 20th September, Mr Stirk had an informal meeting with the Commissioners' Clerk. It seems that by this time, a firm decision had been made to close the line. However, it was explained this would only affect the section between Freeman's Wood and Glasson Dock. The Lancaster end of the branch serving Williamson's and the Ford Quay would remain open for the foreseeable future. It was acknowledged that, because of the pre-existing agreement of 1878, closure would require parliamentary sanction. To this end, the BRB signalled their intention to insert a clause in their forthcoming railway bill. Consequently, in December 1963, the Commissioners were served with a copy of the British Railways Bill 1963–64, clause 22 of which was entitled *Repeal of Obligation to maintain Glasson Dock Railways*. The BRB file relating to the closure of the section of the line to Glasson Dock is held at the National Archive. Correspondence within it indicates that BR had anticipated that the Port Commissioners might petition Parliament against the bill and they were, if necessary, prepared to offer some sort of financial compensation for the loss of the railway. The actual sum would be agreed by negotiation. However, for obvious reasons this willingness was not revealed to the Commissioners.

Unhappily, for the Commissioners, during 1963, activity at the port had diminished significantly. The number of vessels arriving had reduced by around a third and the total tonnage was a little over one-half of figure for 1962. This was primarily due to the ending of the nitrochalk shipments in September 1963 and, whilst ICI had retained the conveyor at Glasson Dock, there was no immediate prospect of a resumption of this traffic. Also, around this time, Williamson's had notified the Commissioners that in future they would be using smaller quantities of linseed oil and china clay in their manufacturing processes and consequently there would be fewer vessels visiting the port carrying these commodities. It is likely that this was because of the firm's gradual shift from linoleum to vinyl (PVC) for their floorcovering range.

It is unfortunate that the ending of the nitrochalk shipments at this critical juncture destroyed the only plausible argument the Commissioners could have put forward for the retention of the line to Glasson Dock. Shortly after being served with a copy of the bill, several the Commissioners made a trip to London to seek legal advice on the closure from the parliamentary solicitor to the Dock and Harbour Authorities Association. However, they were advised that, given all the circumstances, no useful purpose would be served in attempting to seek parliamentary approval for any amendment to the bill. With all this in mind, and perhaps with some reluctance, the Commissioners decided not to pursue the matter any further. By February 1964, the parliamentary bill had had its second reading and an objection had been lodged by Messrs Nairn Williamson's Ltd, the new owners of Lune Mills. It should be explained that, in 1963, James Williamson & Son Ltd and their long-time Scottish rivals, Michael Nairn & Co of Kirkcaldy, merged to form Nairn Williamson Ltd. Although the part of the branch serving Lune Mills was unaffected by the proposed closure, the new company was concerned about the termination of the railway's legal responsibility to maintain the line. There was a fear that with this gone, the remainder of the facilities might be closed without warning at some point in the future. At that time, Nairn Williamson Ltd was a frequent and regular customer of the railway at the Lancaster end of the branch. As well as sending out trains of vans loaded with floorcoverings, the company's coal supplies for their power station still arrived by rail. In fact, figures in the BRB file relating to the closure indicate that, in 1962, traffic at the Lancaster end of the branch remained healthy. During the year, outbound traffic had amounted to 12,050 tons carried in

RCTS Rail Tour to Glasson Dock 20th June 1964

Once it became clear that most of the branch was about to close, the Railway Correspondence and Travel Society organised a final rail tour to Glasson Dock. The passenger accommodation consisted of six goods brake vans, which appear to have been crowded with railway enthusiasts, most of them smartly attired in shirts, ties and jackets. The locomotive was Ivatt 2MT No 46433, travelling tender first from Lancaster Castle Station to Glasson.

Derrick Codling.

On the trip down to the terminus, at least one stop was made *en route*, to enable the participants to take photographs. Once at Glasson, No 46433 utilised the loop at the station to run around the short train in readiness for the return journey. By this time, the rails here had all but disappeared under a carpet of grass and weeds. The locomotive is seen here standing just outside the station in the process of running around its train.

Derrick Codling.

On the return to Lancaster, the locomotive ran smokebox first. Here the train has just crossed the viaduct over the River Conder and is about to pass the site of the former station at Conder Green. It is clear from this and other photographs that at least some effort had been made to spruce up No 46433 for this duty. As well as displaying some signs of attention by the cleaners, the smokebox numberplate had been painted blue and the numbers picked out in white.

Derrick Codling.

4,190 wagons. The corresponding inbound tonnage figure was 49,090, carried in 4,430 wagons. Included in the latter figures was the tonnage of coal coming into Williamson's power station, said to be the most profitable part of the traffic, and amounted to around 46,000 tons *per annum*. The Railways Board was able to assure Nairn Williamson's that given the level of traffic, there was no desire to close the Lancaster end of the branch. Furthermore, BRB was able to provide a written undertaking that, in the event of them deciding to close this section at some point in the future, they would provide twelve months' notice. Following these assurances, Nairn Williamson's withdrew their objection.

The British Railways Act 1964 finally received royal assent on 10th June 1964, thereby sealing the fate of the Glasson Dock–Freeman's Wood section of the branch. Once it became clear that much of the branch was to close, a final railway enthusiast's special was hurriedly organised. On 20th June 1964, Glasson Dock was visited by a rail tour organised by the Railway Correspondence and Travel Society. The passenger accommodation was composed entirely of goods brake vans and the engine was another Ivatt 2MT 2-6-0, this time No 46433, which appears to have been cleaned up specially for the occasion. To add an extra touch of distinction, the numberplate on the smokebox door had been painted blue and the numbers picked out in white. The train stopped at various points along the branch to enable the enthusiasts to take photographs.

On 10th August 1964, the Commissioners received a letter from the Divisional Manager, British Railways London Midland Region, saying that once the statutory requirements had been met and closure had been completed, the railway would begin to remove the redundant track and infrastructure. He proposed a site meeting at Glasson Dock to settle any outstanding points, especially given that the lines owned by the Commissioners at the dock would become isolated following closure. This meeting took place on the 1st September, and the Commissioners' minutes suggest that the railway representatives offered the use of their own contractor to lift any sections of their track that needed removing. However, the Commissioners were content to leave these *in situ* for the moment. In due course, they would arrange with their own contractor to remove the portions of the raised track that might have some scrap value. However, any sections of rail embedded in sets would be left in place. Therefore, from 7th September 1964, all traffic to Glasson Dock ceased. There is a handwritten note in BRB's closure file, made some two years after the event, to the effect that the line was closed on the 1st October 1964. However, this seems to have been an error, especially since the 1st October was a Thursday. In one sense it is academic, as apart from the enthusiast's special in June, it is unlikely that anything had travelled to the terminus since March of that year. The line was truncated at a point one mile and 143 yards from Lancaster No 4 signal box, situated at the junction with the main line. A note in the BRB closure file indicates that, by late October 1964, work was in hand to recover all materials and to demolish the station building at Glasson. By November 1966, the only work remaining was the removal of around 200 scrap sleepers from the site of the station and about 50 telegraph poles at the Lancaster end of the closed section. At this point, the former station master's house was in the process of being sold, and negotiations were under way for the sale of much of the western end of the closed line to Lancashire County Council. However, the remaining part of the branch remained open and in use, owing principally to the continued activity at Nairn Williamson's Lune Mills site and the small amount of traffic to and from the District Civil Engineer's Lune Road workshops. Unfortunately, the traffic from the former was not to last. After three consecutive years of falling profits and a loss in the first half of 1966, a decision was made to restructure the business. In February 1967, work began to transfer the manufacture of all smooth floor coverings in linoleum and vinyl to Kirkcaldy. However, all coated products and pile fabrics would now be manufactured at Lune Mills. The loss of the modern and increasingly popular lines in vinyl was something of a blow to Lancaster and resulted in some redundancies. The transfer to Kirkcaldy took place throughout 1967. From this point onwards, there was a gradual reduction in the amount of rail traffic generated by the Lune Mills site. Local memory recalls that coal trains to the power station ceased in 1967 following a switch to oil for firing the boilers. Similarly, the warehouse, served by the private siding off the Ford Quay line, is said to have closed on 1st March 1968. However, a letter from BR's Area Manager dated 8th September 1969 indicates that at this point, some traffic was still passing to Nairn Williamson Ltd.

The first part of the remaining rail infrastructure to close was the siding leading to the gasworks. A notice dated 30th December 1968 recorded that:

> *The Gas Works siding, situated approximately 500 yards along the siding leading to St George's Quay will be taken out of use pending removal. The hand points leading to the Gas Work will be secured out of use pending removal.*

Given that gas production on the site ceased in around 1960, it is likely that the connection had seen very little use for some years. The remainder of what had constituted the Lancaster Quays goods line was closed to traffic on the 30th June 1969. This now consisted of the sidings onto the quay itself, the connection into the warehouse at Lune Mills, and the long siding running along the rear of the site towards Freeman's Wood. To the end, the railway had maintained two wooden huts on the quay side, which, together, constituted an enquiry office. Shortly afterwards, the connections with the former main running line to Glasson Dock were clipped and padlocked pending removal. This left just the single line from Castle Station, together with the connections to the District Civil Engineer's yard and Williamson's power station. A memo in the BRB's closure file indicates that this final section of the branch was closed to traffic on 4th April 1971. On this date, the facing connection between the down platform line and the remaining stub of the branch was secured out of use pending removal. However, it is likely that all traffic on the branch had ceased at some point before this. This appears to have been carried out as part of the work to remodel the bay platforms at Castle Station, which took place at this time. Subsequently, the connection, together with its associated signalling, was removed. Local memory recalls that the remaining track on the branch was lifted during the Spring and early Summer of 1971.

Postscript

Lancaster County Council did, indeed, purchase much of the western end of former route of the railway. This was turned into a public right of way and today it is readily accessible to both walkers and cyclists. The route can be reached from Lancaster by following St George's Quay and keeping the River Lune on your right. The road continues as a footpath and after following this for around a mile, the old railway route is joined in the vicinity of Aldcliffe. The two routes finally merge into one at what was formerly Aldcliffe level crossing. From here, the cyclist or walker is able to gain a reasonable impression of what it was like to travel as a passenger on the branch up until 1930. Inevitably though, views of the estuary and the surrounding countryside are restricted in many places, owing to the growth of trees and vegetation since closure. The highlight of the journey is perhaps emerging from the wooded car park at Conder Green to pass along the gently curving viaduct over the River Conder for the final run into Glasson Dock itself. Today it is perhaps a little difficult to locate exactly where the station building used to be. However, the stationmaster's house is still there, having long since reverted back to its original function as a private dwelling. The site of the storage sheds and the sawmill are now covered by a car park catering for the significant number of visitors who come to the village, especially at weekends in Summer. The shipyard and graving dock are long gone, and much of the canal basin is now a marina for pleasure craft. Happily though, in spite of all of the vicissitudes, many of which have been documented within these pages, Glasson Dock remains a busy working port and the Lancaster Port Commission is now the Statutory Harbour Authority. Cargoes handled include animal feedstuffs, fertilisers and aggregates and the main port user, Glasson Grain Ltd, has a stevedoring department with a wide range of handling equipment, together with facilities for both indoor and outdoor storage. Vessels of up to around 90 metres long now use the river berths, and the wet dock can accept vessels of up to 85 metres in length. In some ways it's very different from the heady days around 1900 when Glasson Dock was busy with iron ore traffic from Spain. However, in others perhaps, it is reassuringly similar.

Chapter Nine

Working the Branch

FROM AN OPERATIONAL perspective there were, for much of the line's history, two very distinct single line sections of the Glasson Dock branch. By far the longer of the two, left the main line immediately to the north of Lancaster Castle station and provided the route to Glasson Dock. This was the branch's main line, so to speak and, until 1930, it carried both goods traffic and the passenger service. The second shorter section, the Lancaster Quay goods line, also began at the junction with the main line and ran parallel with the Glasson Dock line, to a point just short of an occupational crossing at Freeman's Wood, just a little way beyond Williamson's Lune Mills site. In addition to this, there was a siding running north eastwards from the goods line onto the Ford Quay, situated alongside the River Lune. As the name suggests, the Lancaster Quay goods line only ever carried freight traffic and until well after the passenger service ended in 1930, this was to all intents and purposes operated as a separate entity from the main route to Glasson. Rather than having a scheduled daily goods service, this part of the branch was served by a series of trip workings from Castle Station goods yard, these being carried out by the designated Lancaster shunting engines,

Much of what follows is taken from various timetables and documents generated by the LNWR and its successors. As well as providing passenger timetables for the benefit of the public, the LNWR, in common with other railways, also issued what were known as working timetables (WTT). These show all train and light engine movements as well as providing a great deal of ancillary information relevant to the day-to-day operation of the railway. They were issued solely for the use of the company's employees and consequently, their survival is less common than the public timetables, especially for the years prior to 1900.

The earliest WTT available for the branch dates from October 1884, and this reveals that whilst there were four trains from Glasson Dock to Lancaster at 8-45, 10-00, 1-15 and 5-30, there were only three trains in the opposite direction at 9-15, 12-00 and 4-10. This requires further explanation. The first train of the day down the branch, was a goods working to Glasson Dock, which left Lancaster at 8-00 am and also brought with it the empty branch carriage set. After the locomotive had deposited its wagons on the quays and carried out any shunting, it worked the 8-45 passenger train to Lancaster. In addition to this early morning goods train, the 12 noon passenger service from Lancaster and the 5-30 pm from Glasson Dock were mixed trains, i.e. as well as the usual passenger carriages, a number of wagons might also be attached to the rear. The WTT makes it clear that during the layover periods at Glasson Dock, the engine was to be available to carry out any shunting that might be required at the terminus. A later WTT from March 1887 indicates that apart from some minor revisions to the timing of the trains from Glasson Dock, the arrangements for the goods and passenger services remained the same.

By 1890, the public timetable indicates that there were four passenger trains in each direction with the timings from Lancaster at 8-00, 9-20, 12-00 and 4-15. From Glasson the times were 8-40 10-40, 1-15 and 6-08. There are no details of the goods services of course, but it is likely that the traffic to and from the terminus was handled by the same mix of a single daily dedicated goods train and two mixed passenger services. Also appearing in this timetable is the small station at Condor Green, which had been opened for traffic on the 5th November 1887. The station was only open on Saturdays and the trains were arranged so as to provide the scattered local population with the opportunity of visiting the market in Lancaster. With this in mind, the first three trains on Saturdays from Glasson Dock to Lancaster also called at Conder Green whilst in the opposite direction, the last two trains from Lancaster also called there. Later WTTs reveal that an additional three minutes was added to the arrival times at Glasson Dock and Lancaster to allow for the stop.

The 1899 WTT confirms the frequency of passenger trains down the branch although there are some variations as to timings. Additionally, the 12-05 ex-Lancaster and the final service of the day from Glasson Dock were designated as mixed trains, thus replicating the arrangements pervading in 1884. In addition to the mixed trains, the branch was now served by two dedicated goods trains per day, one in each direction. However, these were conditional, which simply means that they were run only when there was a sufficient amount of traffic to be moved to or from the terminus. The early morning conditional goods train, scheduled to arrive at Glasson Dock at 7-35, would bring full and/or empty wagons and place them on the quay lines, to be loaded or unloaded during the course of the working day. Having deposited the wagons, the engine had to return fairly quickly back to Lancaster Castle Station in order to work the first branch passenger train to Glasson Dock at 8-00am. If the level of traffic to or from the terminus necessitated an afternoon goods train, an engine and guards van departed from Lancaster at 2-55, arriving at Glasson Dock at 3-05. This then gave the train crew

Glasson station on the 11th March 1964. Ivatt 2MT No 46422 halts during shunting operations, while the travelling shunter poses for the camera. The grass-grown nature of the whole railway estate, so much a feature of Glasson Dock by this time, is readily apparent. In the background are the ship-breaking berths operated by Lacmots Ltd. Initially, these provided some traffic in scrap metal for the branch.

Ron Herbert.

and shunter 45 minutes to extract the outgoing wagons from the dock sidings and marshal them for the return trip leaving Glasson Dock at 3-50. It is likely that on the days when there was no traffic to or from the dock, the two mixed trains were sufficient to handle the small amount of goods traffic generated locally.

This increase in the provision of goods services to the terminus is a reflection of the heightened level of shipping activity at the wet dock and the river quay. In the period 1896 to 1905, Glasson Dock was relatively busy with incoming cargoes of Spanish iron ore for Carnforth Ironworks, wood pulp for Cropper's at Burneside and cork for Williamson's. It is possible that at the busiest times, the conditional goods trains on the branch, in themselves, were insufficient to clear the cargoes being discharged and additional trains were run. However, these would not have appeared in the relevant WTT and may have been the subject of special traffic notices; or, perhaps more likely, the decision to run an additional train would have been made on the spot by the District Traffic Superintendent.

The WTTs for 1905 and 1909 reveal a very similar pattern of services with some slight variations in timing. However, the designated mixed train from Lancaster is now the 8-00, the first passenger train of the day to the terminus. This is still worked by the engine off the early morning goods train to Glasson Dock. In 1909, the trains from Lancaster departed at 8-00, 9-20, 12-05 and 4-45. In the opposite direction the trains left Glasson Dock at 8-35, 10-15, 1-15 and 5-40. These timings are very similar to those appearing in the 1890 public timetable. However, the 1909 WTT, which came into effect in October of that year, reveals that at this point all the passenger trains had third class accommodation only. Hitherto, the passenger trains on the branch appear to have been provided with accommodation for all three classes. This feature is not repeated in subsequent timetables from the LNWR period and the matter is discussed in more detail in chapter 11.

By 1912, there had been some changes to the arrangement of the trains calling at the station at Conder Green. Now, all Glasson-bound trains on all days called to set down passengers, providing the guard was notified beforehand at Lancaster. However, in the opposite direction, the arrangements from 1909 still prevailed, with only the first three trains on Saturdays calling at Conder Green. It is perhaps appropriate at this point, to discuss the passenger service at the branch's other intermediate station, Ashton Hall Halt. This never did appear in the public passenger timetables nor is it shown in any of the extant WTTs. It is said that intending passengers flagged down the approaching train and it is presumed that anyone intending to alight at the halt would have had to notify the guard before the commencement of the journey. However, it has not been possible to trace any official reference to this practice. Presumably Mr Starkie, who requested the provision of the halt in 1883, found it convenient to use it from time to time. James Williamson of linoleum fame purchased the Ashton Estate in 1884, but by arrangement, Mr Starkie and his family continued to live at Ashton Hall for some years after the sale. One source states that the halt was transferred to Williamson (by this time Lord Ashton) on the 15th June 1920. However, no evidence has been discovered to suggest that he ever used the facility personally.

At various times, the passenger service on the branch might be reduced due to coal shortages or by the exigencies of war. For example, during the 1912 miners' strike, which ran from the end of February until 6th April, a number of services on the LNWR's Lancaster and Carlisle section were discontinued in an attempt to eke out dwindling coal supplies. These reductions took effect from the 11th March 1912. On the Glasson Dock branch, it was the 9-20 service from Lancaster and the corresponding return service from Glasson that disappeared temporarily from the timetable. However, from the 25th March 1912, the service was further reduced to only two trains per day in each direction. During the Great War, the WTT for 1915 reveals that at this point, there had been no alteration in the number of trains between Lancaster and Glasson Dock. The pattern of service also remained as before with four daily passenger trains in each direction, with the very first and very last trains being mixed. Additionally, there were still two conditional goods trains, one in each direction. However, from the 1st January 1917, there were wholesale reductions in passenger services across the railway network, together with significant increases in passenger fares. The purpose of these measures being to reduce non-essential travel so that more resources could be directed towards the war effort. Details of the reductions on the Glasson Dock Branch do not appear to have survived. However, it is likely that as in 1912, the service was reduced to two daily return trips to the terminus.

In the immediate aftermath of the Great War, the 1919 WTT reveals little change. By this time, any reductions in the level of the passenger service had been made good and we see the usual four trains in each direction on every weekday. However, whilst the first and second trains from Lancaster are now designated as mixed, there is no corresponding mixed service in the opposite direction. The arrangements for the conditional goods trains are almost identical to those pervading before the war, with a morning train from Lancaster arriving at Glasson Dock at 8-15. If an afternoon goods train from Glasson Dock was required, an engine and van would leave Lancaster at 2-20, arriving at its destination ten minutes later. The train left the terminus at 3-50 providing ample time to carry out any shunting of the sidings serving the wet dock and the river quay.

The WTT for 1923, the year in which the LMS came into existence, reveals that by this time, there had been some significant changes to the timetable. The first thing to say is that the mixed trains have disappeared. The passenger train departures from Lancaster were at 7-00, 9-10, 12-10, 5-00 and 6-10. The corresponding departures from Glasson were at 8-25, 10-10, 1-25, 5-25.and 6-45. However, on Saturdays only, there were two additional, later trains, one leaving Castle Station at 7-30 pm and returning from Glasson at 8-00. The other left Castle at 9-00 pm and returned from the terminus at 9-30. This latter return

By 1961, the station building was looking distinctly dilapidated. The reduction of the platform area seems to have given the scene a faintly transatlantic air. This, together with the overgrown track, almost gives the impression of a long abandoned railway depot somewhere in the Americas. However, at this point, Glasson was still a fully operational part of British Railways.

Ron Herbert.

GLASSON DOCK BRANCH (Single Line).

Distance.	WEEK DAYS. DOWN TRAINS	1	2	3	4	5	6	7	8	9	10	11	12	13
		Mixed Train.		C Goods		Mixed Train.	Pas			Pas				
Miles.		a.m.		a.m.		a.m.	p.m.			p.m.				
0	Lancaster ....dep	7 30	...	7 50	...	9 20	12 20	...	...	5 0	...	...	...	...
$4\frac{3}{8}$	Conder Green	...	...	...	...	...	...	...	...	...	...	...	...	...
$4\frac{7}{8}$	Glasson Dock ....arr	7 46	...	8 15	...	9 36	12 36	...	...	5 16	...	...	...	...

Distance.	WEEK DAYS. UP TRAINS	14	15	16	17	18	19	20	21	22	23	24	25	26
		Pas		Pas		Pas				C Goods		Pas		
		a.m.		a.m.		p.m.				p.m.		p.m.		
0	Glasson Dock ....dep	8 25	...	10 15	...	1 5	...	...	...	3 50	...	5 25	...	...
$0\frac{1}{2}$	Conder Green	...	...	...	...	...	...	...	...	...	...	...	...	...
$4\frac{7}{8}$	Lancaster ....arr	8 41	...	10 31	...	1 21	...	...	...	4 15	...	5 41	...	...

Nos. 1, 5, 6, 7, and 9—Call at Conder Green when required to set down passengers. Driver and guard to be instructed at Lancaster; three minutes more time allowed when stop is made.

No. 3—When this train is run, the engine must return at once to Lancaster.

Nos. 14, 16, & 18—On Saturdays only will call at Conder Green to pick up market passengers for Lancaster, and will be allowed 3 minutes more time for the extra stop.

No. 22—When this train is run, the engine, van, and guard will leave Lancaster at 2.20 p.m., arr. Glasson Dock 2.40 p.m.

Goods, &c., Trips—

Lancaster Castle Station, Old Station, and Quay Sidings.

Shunting Engine No.	2		2		3		2				2		3	
			A		S		A							
	a.m.		a.m.		a.m.		p.m.		p.m.		p.m.		p.m.	
Old Station	...	...	9†30	...	...	...	...	...	...	...	8 †0	...	11† 0	...
Castle Station.....................	1 50	..	9 35	...	10 15	...	3 30	...	5 0	...	8 5	...	11 15	...
Quay Sidings	2 0	...		...	10 25	...	3 35	...	5 10	...	...	...	...	...

Shunting Engine No.	2	2.					3		2		3		2	
	A				S								A	
	a.m.		a.m.		p.m.		p.m.		p.m.		p.m.		p.m.	
Quay Sidings	5 0	...	...	...	12 15	...	4 0	...	...	...	8 0	...	...	...
Castle Station	5 10	...	7 30	...	12 25	...	4 10	...	4 45	...	8 10	...	8 30	...
Old Station		...	7 45	...	...	...		...	5 0	...		...	8 40	...

A—Light Engines. †—Lancaster Junction.

Midland Trains run between Lancaster (Green Ayre) Mid. and Castle Stations as under:—

Distance.	WEEK DAYS. UP.	Min'rl M	Min'rl		Min'rl			† See note Min'rl S	Min'rl SO		SUNDAY.	Min'rl	
Chain		a. m.	a. m.		p. m.			p. m.	p. m.			a. m.	
...	Green Ayre Station ....dep.	3 15	10 20	...	2 40	...	...	7 35	8 24	...		1 30	...
48	Castle Station.................... arr.	3 20	10 25	...	2 45	...	...	7 40	8 30	...		1 35	...

Distance.	WEEK DAYS. DOWN.	Min'rl M	Goods		Min'rl			† See note. Min'rl S	Goods SO		SUNDAY.	Min'rl	
												A	
		a. m.	a. m.		p. m.			p. m.	p. m.			a. m.	
...	Castle Station.....................dep.	4 0	10 55	...	3 25	...	...	7 55	8 50	...		2 0	...
48	Green Ayre Station .... arr.	4 5	11 0	...	3 30	...	...	8 0	8 55	...		2 5	...

†—These trips are limited to 15 wagons.

A—Travels over Loop Line from No. 2 Box to Midland Junction.

A view across the level crossing looking towards the canal basin. Prominent on the left is the lever frame, which for many years operated the signals protecting the level crossing and also the gate lock. Behind this can be seen a corner of the former station master's house. On the right is the curved roof of one of the twin storage sheds constructed by the LNWR shortly after the opening of the line.

CRA, Les Brough collection.

service also ran on Wednesdays. Furthermore, on Wednesdays and Saturdays only there was an additional service leaving Lancaster at 2-15 and returning from Glasson at 3-05. All this represents a significant enhancement to the passenger timetable as it was prior to the Great War, with an additional later train on weekdays and extra trains Wednesday and Saturday which were Lancaster's market days. The later services on Saturday would provide the inhabitants of Glasson with the opportunity to attend the cinema or similar entertainment, or perhaps allow them to linger for a little longer in one of Lancaster's many pubs. The 1923 WTT also reveals that, by this time, the last three daily return services were operated by an ex-LNWR steam railmotor. It is not known when these were first utilised on the Lancaster–Glasson Dock services, but the available documentation suggests that it was not before 1921. However, in 1923, the first two return trips each day, the Wednesdays and Saturdays only services and the later trains on Saturdays, were all still operated using a conventional locomotive and carriages. By 1923, the service provided for Conder Green had also improved somewhat. Now, on weekdays all trains in both directions called at the station. However, the trains from Lancaster would only call to set down passengers and in the opposite direction, they would only pick up those intending to travel to Lancaster. It was therefore not possible to book a ticket for travel between Conder Green and Glasson.

Until we reach the 1920s period, there is very little information relating to special or excursion trains on the branch. In 1884, shortly after the opening of the branch, it is known that the LNWR ran a special train to take passengers to watch the launch of a vessel from Nicholson's shipyard. Whether this was repeated for subsequent launches is unclear. Certainly, a photograph of the launch of the *Argus* in 1906, the last large ship constructed by Nicholson's shipyard, suggests that the number of spectators was insufficient to merit the provision of a special train. Mr Gilmour, the last station master at Glasson, recalled that every year there were one or two Sunday school trips from different Lancaster schools. However, the numbers presenting themselves for travel were such that the normal branch set of carriages was sufficient for the purpose. Although no confirmatory evidence has been uncovered, it is likely that this pattern repeated itself over many years from the opening of the branch. Mr Gilmour also recalls a large excursion that visited Glasson on the final day of the passenger service. This was organised by the Salvation Army and was run from a town in south Lancashire. The train comprised eight carriages and was hauled by two locomotives. It was recalled that most of the passengers left the train at Conder Green and headed straight for the shore. Given the limited facilities at Glasson station, the length of the train created some problems and it proved necessary to divide it into two, to enable the locomotives to run around the carriages. The re-formed train was stabled on the line to the river quay to await the return journey.

An extract from an LNWR working timetable effective from May 1919. This shows the passenger and goods service between Lancaster and Glasson, the latter being accommodated by a combination of mixed passenger trains and conditional goods trains. Below this are details of the Lancaster goods trip workings which divided their duties between the quay goods line and the sidings at the Old Station Yard. Courtesy Richard Foster.

On the goods side of operations, in 1923, there was an early morning conditional service that left Lancaster at 7-50, arriving at Glasson Dock at 8-15. It is known that at this point the early morning services on the branch were covered by one of the engines allocated to the Lancaster shunting and banking turns. By 1924, this particular duty was in the hands of an ex-Lancashire and Yorkshire Railway Class 22 engine. These were 0-6-2 passenger tank engines dating from the early 1880s. Latterly the LYR had utilised them as pilots at its largest passenger stations. However, following the Grouping in 1923, some of them seem to have moved northwards, being allocated to Garstang Town shed on the old Garstang and Knott End Railway and to the ex-LNWR shed at Lancaster. The WTT suggests that these ex-LYR 0-6-2 tanks worked the first passenger train of the day from Lancaster and after leaving the carriages at the terminus would, if necessary, depart smartly back to Lancaster, to work the conditional goods service back to Glasson Dock. This was scheduled to arrive at the terminus at 8-15. There appears to have been barely enough time to deposit any wagons on the quayside, before the engine had to return yet again to Lancaster with the 8-25 passenger service arriving at Castle Station at 8-40. The ten minutes available for shunting at Glasson Dock does make this scenario seem slightly implausible. Shunting the quayside sidings was far from straightforward. However, the relevant shunting engine roster, allocates the time on these duties quite precisely between 7-00 am and 8-40, which fits perfectly with the initial departure from and the final arrival back in Lancaster. In the afternoons there was a regular goods train departing from Lancaster at 2-35 and arrived at Glasson Dock at 3-00. After shunting for 50 minutes the return service departed at 3-50.

During the prolonged miners' strike, which followed the General Strike of 1926, the passenger service was reduced to three trains per day leaving Lancaster at 7-10am, 9-10 and 5-00pm. These returned from Glasson at 8-25am, 10-10 and 5-25pm. In addition to this there was a Saturdays-only service leaving Lancaster at 11-55 and returning at 12-20.

Recalling the timetable as it was in the late 1920s, Mr Gilmour stated that the basic pattern of four daily passenger trains was a long standing one and that the extra trains on Saturdays were only introduced after the Great War as a concession to the travelling public. This is certainly borne out by the surviving timetables.

However, by the time Mr Gilmour took over the station in October 1927, it would appear that the additional trains on Wednesdays had been deleted from the timetable, as he makes no mention of them. This is confirmed by the LMS passenger WTT, effective from 23rd September 1929, the last full year of the passenger service. This shows a slightly reduced pattern of services to that encountered in 1923, with a basic service of only four daily trains with three additional services on Saturdays.

In relation to the goods service on the branch in the late 1920s, Mr Gilmour provides a slightly different version of events to that revealed in the 1923 WTT, mentioning only a single conditional daily goods train leaving Lancaster at 2.-30pm. During the course of a telephone consultation with Carnforth Control Office each morning, a decision would be made as to whether the train was to run. It seems, therefore, that the early morning conditional goods train had disappeared from the timetable by this time. This certainly fits with what is known about the amount of rail traffic generated by vessels visiting Glasson Dock, which by this time was at a low level. However, Gilmour goes on to explain that it was permitted to attach a single wagon to the 9-10 or 12-10 passenger trains from Lancaster and to the corresponding return trains from Glasson Dock. Until around 1928–29 these were worked by the ex-LNWR railmotor. In this way, when goods traffic was light, it could be hauled by these passenger trains and the afternoon goods service could be dispensed with. In terms of passenger numbers, these two services are likely to have been the least well patronised. It may be that these were deliberately chosen so as not to tax the railmotor too much in the event that a wagon was attached to the rear. Mr Gilmour recalls that it was an interesting sight to see the railmotor shunting the dock lines in order to retrieve its single loaded wagon.

The LMS goods WTT from September 1929 essentially confirms Mr Gilmour's recollections and shows a single daily afternoon goods train running between Lancaster and Glasson Dock. It arrived at the terminus at 3-00 pm and departed for Lancaster 50 minutes later, allowing a reasonable amount of time to shunt the sidings serving the wet dock and the river quay. However, this train did not run on Saturdays and there is no indication in the timetable that its running was conditional on their being sufficient traffic available.

Having considered the workings between Lancaster and Glasson Dock up to the cessation of the passenger service in 1930, we will now look at some of operational features of the sidings on the quay side at the terminus. These were always awkward to shunt, consisting merely of several dead-end roads with no facility for a locomotive to run around its train. Consequently, drivers of goods trains would utilise the short loop at the passenger station for this purpose, splitting the train into two where necessary. With this accomplished, the rakes of wagons would be pushed into the sidings alongside the wet dock or the river quay. On the former, this presented difficulties for the locomotive driver as his forward view was obstructed by the large storage sheds and the sawmill, which lay alongside the canal basin. The position was further complicated by the fact that, at the end of the canal basin, the line had to cross an ungated road before gaining the quayside proper. Instructions, dating from 1919, stipulated that no more than 20 wagons were to be propelled from the station to the dock at any one time. The guard or the shunter was to walk in front of and alongside the first wagon and at the corner of the canal shed was to satisfy himself that the way was clear to the trap points close to where the ungated road crossed the dock line. Just before the trap points, he was to halt the train and ensure that the road was clear of traffic, before signalling the driver to move forward. Part way along the canal basin, there was a fixed hand crane used from time to time to load or unload vessels in the basin. If this was in use at the time of the shunt, the guard or the shunter had to come to an understanding with the men in charge before proceeding.

On the quayside itself, there were originally two parallel sidings on both the dock and the river quaysides, with a third siding being added alongside the wet dock in 1903. Additionally, the two groups of lines serving the dock and the river quay respectively, were connected on the pier head *via* a wagon turntable, which added to the operational flexibility. Once wagons had been placed on the quayside, they could be loaded or unloaded alongside the ship and then, later on, retrieved and returned to Lancaster. The two mobile steam cranes at Glasson Dock — the property of the Port Commissioners — were originally fitted with what were described as steam winches. These were intended to be used to move individual wagons around on the quay lines. However, it is not known to what extent this facility was used by the LNWR, or indeed if it ever was.

We turn now to the operation on the Lancaster Quay goods line. At its western or Freeman's Wood end, this served Williamson's Lune Mills site and in addition to this, a siding ran onto the Ford Quay and provided rail access for Williamson's St George's Works and Lancaster Corporation's gasworks. This siding also provided rail facilities for vessels loading or discharging at the Ford Quay. It also served the collection of sheds on the quay, which, on the eve of the Great War, were occupied by several businesses whose common denominator was that they all brought in materials by sea.

The earliest document providing information about the working of the quay goods line is the WTT from October 1884, which indicates that, at this point, the engine allocated to the goods and passenger trains to Glasson Dock also handled all the traffic on the quay goods line. The branch engine came off shed at 7-00 am and its first task was to work one return trip to what is described as the Lancaster Quay Siding. Having accomplished this and returned to Castle Station, it took the early morning goods train to Glasson, which left Lancaster at 7-45. The WTT makes it clear that during the subsequent layovers at Lancaster,

This image is not of the best quality, but it is the only pre-1923 view showing a locomotive shunting the Commissioners' sidings alongside the wet dock. Apart from the Coal Tank, there is an interesting array of private-owner coal wagons from the Wigan area. Also to be seen is a fitted LYR 12-ton van with what appears to be its smaller, unfitted counterpart, an LNWR goods brake van and an open wagon from the London and South Western Railway.

Courtesy Lancaster City Museums.

On the 2nd March 1962, Ivatt 2-6-0 2MT No 46422 slowly propels a train of open wagons past the canal basin, across the public road and onto the Commissioners' sidings alongside the wet dock. In the distance, the shunter can be seen supervising the movement. This is one of the special trains from Heysham Moss Sidings to Glasson, conveying nitrochalk from the ICI plant at Heysham.

Ron Herbert.

the engine was to be available to transfer wagons between Castle Station and the quay goods line, as required, and carry out any shunting. Having worked the final train of the day from Glasson Dock to Lancaster, the locomotive was scheduled to make a final foray down the quay goods line to clear out any outbound traffic before returning to the engine shed. These arrangements are confirmed in a later WTT dating from March 1887.

The WTT effective from January 1899, shows that by this time, the quay goods line no longer featured in the timetable for the route between Lancaster and Glasson Dock. Although there is no specific confirmatory evidence, these goods-only duties must have been carried out by means of a series of trip workings from Castle station, which were handled by one of the designated Lancaster shunting engines. This division of responsibilities reflects the large increase in traffic on the branch during the 1890s. As well as a significant rise in the cargoes of iron ore, wood pulp and cork being discharged at Glasson Dock, the 1890s saw a huge expansion of Williamson's Lune Mills site with a consequent increase in the level of traffic at the Lancaster end of the branch.

This mode of operation is confirmed by the WTT effective from October 1909. By this time, the duties on the quay goods line were split between two trip working turns, which were handled by two separate engines. These divided their time, running between Castle Yard and the quay goods line or the Old Station Yard, the latter being a little to the south of the Castle Station complex. As far as the quay line was concerned, the working day began at 1-50 am with a trip from Castle Yard with wagons destined for the different sidings on the branch. After shunting for three hours, the light engine returned to Castle Station. The next foray down the goods line was at 10-15am, and the engine brought down more wagons and then shunted the sidings before returning up to the main line at 11-40am. At 2-45pm, a light engine ran down the goods branch, returning at 3-15 with outbound wagons. The limited amount of time spent shunting suggests that the purpose of this trip was to clear one of the sidings of outgoing traffic, possibly the line running down the rear of the Lune Works site and utilised by Williamson's. The final trip of the day left Castle Yard at 4-30 and, after shunting the sidings for over three hours, returned with wagons at 8-00pm.

All of this suggests that by 1909 there was a considerable amount of activity on this part of the branch, generated principally by Williamson's Lune Mills and St George's Works but also by the Corporation's gasworks. Certainly, all the available evidence indicates that the quay goods line was the conduit for virtually all of the outbound traffic from both of Williamson's sites alongside the river. In terms of inbound traffic, we have already seen that apart from a break between 1932 and 1949, the coal for both of these sites was brought in by rail. For other commodities, principally the raw materials used in the manufacture of linoleum and floorcloth, the picture is more varied. As a general rule of thumb, once the cork boats defected to Heysham, most of Williamson's raw materials came upriver in small coasting vessels and were discharged at either the Ford or the New Quays. However, from time to time, the ship in question would be too large to come upriver to Lancaster and, therefore, would have to discharge its cargo at Glasson Dock. When this happened, the cargo would make the final part of its journey by rail. By the eve of the Great War, however, Williamson's possessed a fleet of motor lorries and it becomes increasingly possible that some of the cargoes discharged at Glasson Dock made their way up to Lancaster *via* this mode of transport.

The sidings on the quay goods line must have been quite difficult to shunt, given that until the Second World War, there was no facility for a locomotive to run around its train and reverse direction. On the Ford Quay there were two dead-end sidings, with a third siding curving sharply into the gasworks. At the opposite end of the goods line, Lune Mills was served by a single long siding, running the entire length of the site and extending down towards Freeman's Wood. This was, in effect, a somewhat extended head shunt for the sidings on the Ford Quay. Given the limitations of this layout, it was necessary to marshal wagons in a certain order according to their destination before any train ventured onto the goods line. Things were made a little easier in 1910, when a crossover was put in between the running line to Glasson Dock and the goods line, at a point on the latter, just to the west of the junction with the line to the Ford Quay. This enabled a train of wagons or a light engine from Lancaster to access the quay goods line *via* the Glasson Dock route. This might be useful where the upper part of the goods line was already occupied by standing wagons, or if a second engine was required to assist with the shunting and the train staff was already in the possession of the driver of the first engine. It also made it possible for goods trains from Glasson Dock to run directly onto the quay goods line without first having to run up into Castle Station in order to reverse direction.

Because of the gradient of 1 in 50 down from Castle Station, the inward trip was always carried out with the engine in front and a brake van at the rear. Depending on the weight of the train, at least some of the wagon brakes would be pinned down for the duration of the descent. Given what was about to take place, it was essential that the rake of wagons and its attendant brake van were completely under the control of the guard and the shunter. The train was usually composed of two blocks of wagons, this being arranged when it was assembled in the yard at Castle Station. At the front would be those destined for the Freeman's Wood end of the quay branch, essentially traffic for the Lune Mills site. At the rear would be the wagons for the Ford Quay including those for the gasworks and Williamson's St George's Works. On the approach to the junction for the line to the Ford Quay, the gradient was 1 in

60 and gravity was utilised in order to divide the train into its two parts. The entire train was halted on the gradient and with the wagon brakes on and the brake in the guard's van screwed down hard, the engine was detached. It ran forward and then reversed, to run a short distance onto the Ford Quay line proper. With the route down to the Freeman's Wood end of the branch now clear, the front half of the train was uncoupled from the rest and allowed to run by gravity to its destination, with the shunter running alongside the moving wagons and making judicious applications of the wagon brakes to ensure a controlled stop in the right place. With this accomplished, the engine was re-attached to the remainder of train and the brake van uncoupled. After running forward, and then reversing again, the engine propelled the remainder of the wagons onto the Ford Quay itself. With the way clear, the brake van was allowed to run down to the Freeman's Wood end of the line in readiness for the return journey. The points for the junction between the line down towards Freeman's Wood and the Ford Quay line were always operated from an adjacent hand lever in order to facilitate these shunting movements. The section of the quay goods line running across the New Quay Road and onto the Ford Quay, was in fact a little operational enclave all of its own. Access onto the New Quay Road crossing was protected by a gate normally kept shut and padlocked. The key for this was held in Lancaster No 4 box and was handed to the shunter or goods guard when required. The gate was opened to allow access onto the quay and then kept closed and locked until the return trip was made. Once on the Ford Quay itself, train movements were carried out in what was essentially a public area and therefore extra vigilance was required. All movements were to be carried out at walking pace with the guard or shunter accompanying the train, walking alongside the first wagon and using his whistle to communicate with the driver and to warn the public that shunting was in progress.

Once the shunting on the quay was completed, the engine would draw any outbound wagons out on to the branch and then propel them a short distance up the gradient. From here they would be allowed to run back towards the Freeman's Wood end of the branch, while the engine removed itself briefly onto the Ford Quay line. From this point it was simply a matter of coupling up the engine, the wagons and the brake van and taking the whole train up to Castle Yard. However, the adverse gradient also imposed some restrictions on the weight of trains returning to the main line. If starting from the Freeman's Wood end of the branch with the distant signal for the main line junction off, the maximum weight, inclusive of the brake van was restricted to 300 tons for mineral traffic and 250 tons for goods. However, if starting from the outer home signal (i.e. closer to the main line junction), these limits were reduced to 240 and 200 tons respectively. The *rationale* behind this distinction was that a train starting from the Freeman's Wood end under clear signals would be able to achieve some speed and momentum before it tackled the adverse gradient.

This reconstruction of events mirrors the instructions contained in the 1919 Appendix to the WTT. However, the details in the 1905 version of this document are essentially the same, suggesting that this method of gravity shunting on the quay goods lines was a long established one. These instructions are not, however, the full story of the operation of this section of the branch. They do not explain, for example, how loaded wagons from the Freeman's Wood end of the siding were extracted, given that there would need to be a brake van at the rear, for the ascent to Castle station. However, this clearly *was* accomplished and on a daily basis. Doubtless everything came off successfully with the application of a mixture of experience, ingenuity and forward planning. Those responsible for the dispatch of outgoing traffic at Williamson's would have needed to work closely with the railway authorities to ensure that: (a) there was a sufficient supply of empty wagons, and (b) loaded wagons were removed from the sidings as quickly as possible. In 1947, following the construction by the LMS of a private siding into the dispatch warehouse at Lune Mills, Williamson's employed their own printed forms for the daily requisitioning of empty vans from the railway. These forms were signed by the warehouse manager and countersigned by the production director, emphasising the importance of having the right number of wagons in the right place at the right time. It is possible that something similar was employed in earlier LMS days and perhaps even prior to that.

It is the 2nd March 1962 and Ivatt 2-6-0 No 46422 is shunting a special train of nitrochalk, which it has brought from Heysham Moss sidings. The wagons have been divided into two rakes so that the short loop at the station can be utilised to enable the engine to run around the train. It appears that the nearest rake has already been unloaded as the wagon sheets have been removed. The second rake (still sheeted over) is about to be pushed onto the siding alongside the wet dock. Ron Herbert.

Having considered the operation of the quay goods lines in some detail, it is worthwhile taking a further look at the relevant trip workings as they were in early LMS days. The following details are taken from a document dated 22nd September 1924, with the resounding title of *Carnforth Control Area Local Trip and Shunting Engine Workings*. At this time, there were three shunting and banking engine turns at Lancaster and turns one and two, shared between them the four daily forays onto the quay goods lines, transferring wagons between the branch and Castle Yard and carrying out whatever shunting was required. In between these duties, the engines also shunted the sidings at the Old Station Yard and at Castle Yard. The designated engine for turn number one was an ex-Lancashire and Yorkshire Railway Class 22 engine. The use of these on the early morning trains to Glasson Dock has already been noted. Turn number two was the province of an ex-LNWR 18 Inch goods tender engine, more commonly known as Cauliflowers. In their day, these had been regarded as express goods engines but by the 1920s they had been superseded by the larger and more powerful 0-8-0s and relegated to secondary duties.

The WTT from September 1929 reveals that by this time there were three daily trip workings down the quay goods line, although the first one of the day did not run on Mondays. The first and the last of these spent between two and three hours on this part of the branch so both trips must have involved a significant amount of shunting. However, there was also a single trip carried out in the early hours of Sunday morning leaving Castle Station at 2-00 am and returning at 5-10 am. This appears to be the only instance of a regular, scheduled working down the branch on the Sabbath.

By September 1938, the daily trip workings had been reduced to two. However, the daily goods train to Glasson Dock also visited the quay goods line during the course of its return to Lancaster. On the outward journey, this train was also scheduled to visit the district civil engineer's yard presumably for shunting purposes. In both directions, it also paused for five minutes at the level crossing at Aldcliffe although the reason for this is unclear. At this time, crossing keeper's houses in remote areas were sometimes not connected to a mains water supply. Instead, it was the practice for a goods train to drop off containers of water and to collect the empties. However, the two visits each day are more difficult to account for. By 1938, there was a further reduction in the level of service on the branch insofar as the daily goods train to Glasson Dock no longer ran on Saturdays.

The WTT for February 1940 reveals an identical pattern to that encountered in 1938. However, by 1942, it appears that the daily working down the branch to Glasson Dock, as well as the trips down the quay goods line, were operated entirely by the designated Lancaster shunting engines. Traffic to Glasson Dock itself was now in the hands of a single daily trip working, which arrived there at 1-50, shunted for 30 minutes and then returned up the branch, calling at the sidings on the quay goods line on the way. Consequently, it did not regain Lancaster Castle Yard until 3-27. As in 1938, this working did not run on Saturdays. The quay goods line itself was also served by two afternoon trip workings from Castle Yard. The first arrived at the sidings at 4-45 pm and shunted until 7-40 when it retraced its route to the main line. The later working arrived at 8-55 and shunted until 10pm. Given that there was a war on, 1942 was far from a normal year, however, the general arrangements for the traffic to and from Glasson Dock and on the quay goods line appear to mirror those existing prior to the commencement of hostilities. By 1943, Williamson's were significantly involved in war work and were making full use of the additional sidings put in at the rear of Lune Mills at the behest of the War Office. These appear to have been utilised for the reception of raw materials connected with this work and for the dispatch of finished products. However, the arrangements detailed above predate this. It would be interesting to examine a WTT from a slightly later time, say 1944, to see what effect, if any, the increased traffic associated with this war work was having on the operation of the branch.

By May 1949, the branch as a whole was served by two local trip working turns and the designated engine in each case was a standard 4F 0-6-0 of which Lancaster's Green Ayre shed had several examples. The earlier of the two turns arrived at the quay sidings at 10-10 am and then shunted until 12-55. However, during this period, the engine would, if required, break off, to make a trip to Glasson Dock to drop off or pick up any wagons. Arrival at Glasson was scheduled for 10-45 and the enginemen were allowed half an hour for shunting, before returning up the branch to the quay sidings. In addition to this, there was a later trip working turn that arrived at the quay sidings at 5-00 pm and then shunted there until 7-45, when it returned to Castle Station. After depositing any wagons in Castle Yard and carrying out further shunting there, the engine made a final foray down to the quay sidings, arriving there at 9-15. It remained there until 11-15, at which point it returned for the final time to Castle Station. On Saturdays, the arrangements were slightly different in so far as there was no late evening second trip down to the quay sidings. By September 1955, the arrangement of the trip workings was almost identical, with only slight alterations to some of the timings.

The BR Freight WTT for September 1961 reveals that, by this time, the quay goods line was served by a late morning and an evening return working. In both instances, a significant amount of time was spent shunting the sidings. In addition to these trains, there was a late afternoon return service that shunted for just under an hour. However, this only ran on Mondays, Wednesdays and Fridays. Glasson Dock itself was served by the late morning service working that, if required, would also travel down the branch to the terminus. Given the very limited amount of rail traffic generated by the dock and the surrounding industries by this time, this is likely to have been an infrequent occurrence. By September 1961, there were no trains on the branch on Saturdays, but the additional service on Mondays and Fridays would have kept the accumulation of wagons for outward dispatch to a minimum. The BR Freight WTT for 1962–63 indicates that on the eve of the closure of the branch west of Freeman's Wood, the operational arrangements were the same.

As we have seen, 1962 marked the commencement of the nitrochalk shipments from the ICI plant at Heysham and the special trains ran until September 1963. It is not possible to say how frequently these workings took place, however, it is known that during August 1963, 1,010 tons of this substance were taken by rail to be loaded into vessels at the dock, suggesting two or possibly three trains during that period. However, as these were run as specials between Heysham Moss Sidings and Glasson Dock they do not appear in the WTTs.

By April 1966, traffic on what remained of the branch was in the hands of a daily morning trip working hauled by an ex-LMS 0-6-0 tank engine from Green Ayre shed. This transferred wagons to and from Castle Yard and shunted the quay sidings between 10-23 and 12-30.

Chapter Ten

Signalling and Train Control

AS WE HAVE SEEN, for much of the line's history, there were, in effect, two Glasson Dock branches, the main running line from Lancaster to Glasson Dock and then the much shorter Lancaster Quay goods line. Until 1930, the line to Glasson Dock carried both goods and passenger trains, whereas the quay goods line was always restricted to freight traffic. Where passenger trains were run, the Board of Trade demanded a significantly higher standard of safety and this is reflected in the differing provision of signalling and train control on the two sections.

Amongst the papers relating to the Board of Trade's inspection of the line in June 1883, is a written undertaking from the LNWR to the effect that the passenger line from the junction at Castle Station to Glasson Dock would be worked using the combined absolute block telegraph and train staff system. Earlier, in January 1883, an LNWR Traffic Committee minute had helpfully provided a summary of what was to be installed on the branch in connection with the former:

A two-wire single line block telegraph to be provided for working the new Glasson Branch between Lancaster Midland Junction Box and Glasson Station with Indicators at Aldcliffe Bond Level Crossing and Stoddart Bottom Level Crossing, together with a single needle speaking circuit between the Lancaster Telegraph Office and Glasson Station and Glasson Dock. Estimate £310.

It is convenient at this point to mention that when the main line junction for the branch was constructed, a new signal box was provided, which controlled both the new junction and the existing junction with the Midland Railway's branch from their Green Ayre Station. This was then referred to as Lancaster No 4, or slightly confusingly, Midland Junction No 4. In 1901, this new box was itself replaced by a new structure as part of the rebuilding of Castle Station, which was completed in 1902.

The combination of the block telegraph and train staff systems provided two distinct controls over the trains on the passenger line. Under the former, the line was divided into sections, each with its own signal box. Only one train was allowed into any one section at a time, and trains were passed from one section to the other by the signalmen using a system of bell codes. The two block posts on the branch were at Lancaster No 4 signal box, situated at the junction of the branch, and at Glasson Station. The whole of the route between Lancaster and Glasson, therefore, constituted a single block section. It should be explained that there was never a physical signal box as such at Glasson. However, the ground frames, together with the telegraph instruments situated in the booking office within the station building, provided the necessary operational capability. In addition to the block telegraph, the trains on the Lancaster–Glasson section were controlled by the issue of a train staff and ticket. At the commencement of the journey from Lancaster, the staff would be handed to the driver and this would constitute his authority to proceed down the branch to Glasson. It was a simple and effective system. There was only one staff, and if the driver didn't have possession of it, he couldn't enter the section. However, there might be rare occasions when two trains needed to be at the terminus at the same time, for example, where a special train was required to clear a cargo being discharged at the port. In this scenario, the driver of first train down the branch would be shown the staff but would be handed a ticket as his authority to proceed down the branch. The Lancaster–Glasson block section probably always ended just outside the station at the latter place. This was certainly the case in later days. Therefore, once the first train was standing at the platform, or had moved onto the lines leading to the quay side, it was effectively out of the block telegraph section, and instead, under the control of the station signals or the shunter. With the first train safely tucked away, the signalman at Glasson would notify his counterpart at Lancaster that the block section was clear. This provided the latter with the authority to hand the staff to the driver of the next train. With both trains at the terminus and the signalman in possession of both the ticket and the train staff, the process would be repeated, with the driver of the first train back to Lancaster being shown the staff and then handed a fresh ticket. Once this first train had arrived safely at Castle Station, Lancaster No 4 box would send the train out-of-section signal to Glasson, whereupon the driver of the second train would be handed the staff as his authority to proceed up the branch. Turning now to the Lancaster Quay goods line. As this was a freight-only line, the installation of the block telegraph system was not required by the Board of Trade. However, all traffic was still controlled by the train staff system. Consequently, Lancaster No 4 box had two different staffs to issue, one for the route to Glasson Dock and the other for the Lancaster Quay goods line. To minimise the possibility of confusing the two, the staff for the former was square in section and coloured red and the one for the goods line was round in section and coloured blue.

In terms of the provision of signals on the branch, the earliest record appears to be an LNWR legal map of the whole branch held at Lancashire Archives. This shows the position of each post and the configuration of the arms. It is undated, but annotations suggest that it depicts the branch as it was in the early 1890s. Helpfully, the information on the LNWR's map can be cross-checked with the first edition of the 25-inch Ordnance Survey dating from 1890, which shows the position of individual signal posts. At Glasson, there were distant, home and starting signals. In addition to these, there was a signal protecting the junction from the river quay siding onto the loop at the station, together with a double signal (i.e. two arms on one post) protecting the level crossing immediately to the west of the station. It seems that at the beginning, these signals and the relevant points were controlled from a single open ground frame situated on its

An excellent view of the two ground frames that controlled access to the sidings serving Williamson's power station. These appear to have been originally installed early in 1943 as part of the work to create a fan of four sidings for the War Department. The sidings were re-purposed in 1946, first of all for the supply of materials for the construction of the power station and then later to provide the site with coal. The lock into which the key on the end of the train staff was inserted can be seen at the bottom of the nearer right-hand lever.

CRA, Rev J Jackson Collection.

This illuminated sign, situated on the approaches to Glasson station, marked the end of the train staff section for the branch. It is likely that this dates from the period shortly after the cessation of the passenger service in 1930. The wording on the sign is reproduced on the diagram covering the later signalling arrangements at Glasson Dock.

CRA, Rev J Jackson Collection.

own raised structure, just off the Lancaster end of the station platform. This is confirmed by both the LNWR legal map and the 1890 Ordnance Survey. However, by around 1907, there appears to have been a second ground frame, situated directly behind the buffer stop at the end of the line serving the platform. This controlled the signalling associated with the level crossing gates and the gate lock. The two frames extant at Glasson were manufactured in 1907 and had consecutive numbers, suggesting perhaps, that they were both installed together, one of them replacing the original frame from 1883.

Between Glasson's distant signal (situated a little to the west of the Conder viaduct) and the junction for the line onto St George's Quay, the only signals were double ones protecting the manned level crossings at Conder Green and Aldcliffe. For reasons that are unclear, there was no corresponding provision at the line's intermediate, manned crossing at Stodday. This is slightly puzzling, because at the time of the Board of Trade inspection in 1883, the inspecting officer, Colonel Rich, instructed that the crossing should be protected by a crossbar signal. However, both the LNWR legal map and the 1890 Ordnance Survey seem to confirm that this was not provided. Colonel Rich's report provides two further minor mysteries. He states that there were authorised crossings of public roads at 1 mile 6 chains and 2 miles 61 chains. The latter location is the manned crossing at Stodday. However, the former location relates to an unmanned crossing for the footpath known as Freeman's Wood, which was definitely not a public road. It appears that the Colonel confused this location with the public road crossing at Aldcliffe, a little further westward and the nearest manned crossing to Lancaster. In his report, Colonel Rich makes no mention of the manned crossing at Conder Green. It seems unlikely that this was a mere oversight and it may be that this was not in place at the time of the Board of Trade inspection. However, the relevant crossing keeper's cottage is known to have been *in situ* by August of 1884.

Close to the junction with the line to St George's Quay, there was a bracket post carrying distant signals for the main line junction for both the Glasson Dock line and the quay goods line. In addition to this, there was a single signal controlling the exit from the line down to St George's Quay.

It is likely that some, if not all of the original signals on the branch were of the older style, slotted-post variety. In 1883, the LNWR introduced a new standard design of signal, which, in some locations, lasted well into BR days. The new design was brought into general use for new work in April of that year but by this time, the line to Glasson Dock appears to have been virtually complete. Indeed, the final inspection for passenger traffic by the Board of Trade took place in the June of that year. Certainly, the arrangements for the new junction at Lancaster Castle station were inspected as early as January of 1883, strongly suggesting that the work here was completed using the older style slotted signals. At some point in LNWR days, the original signalling was upgraded and the old, slotted posts were replaced with the newer standard pattern. It is likely that this work was carried out piecemeal over a number of years. The double signals protecting the level crossings at Glasson, Conder Green and Aldcliffe were replaced with separate signal posts on either side of the crossing. These new signals were interlocked with the level crossing gates. The crossing keepers' houses at Stodday and Aldcliffe, from the outset, were fitted with block telegraph indicators to show the state of the section and whether a train was present. It is highly likely that the third crossing keeper's house at Conder Green was similarly fitted when it was constructed shortly after the opening of the line for passengers.

We have already looked at some of the operational idiosyncrasies of the Lancaster Quay goods line. However, there are some aspects of the train control arrangements for both the goods line and the Lancaster end of the Glasson Dock route that invite closer study. As before, the foregoing details are derived from the instructions contained in the 1919 Appendix to the WTT.

The points for the permanent way works siding were controlled by a ground frame, which was released by the key in the end of the Glasson Dock branch train staff. Therefore, the driver of any locomotive wishing to access the sidings would need to be in possession of this. If there was a train already down the branch heading towards Glasson Dock (and in possession of a ticket), the staff gave the driver of the second train authority to work down the branch only as far as the permanent way siding. In compliance with the rules governing the working of the staff and ticket system, the signalman at Lancaster No 4 box could not hand the staff to the driver until the train out-of-section signal had been received from Glasson. However, it was not deemed necessary to signal the second train to Glasson Dock on the block telegraph. If it was necessary for a locomotive to remain within the permanent way siding to carry out shunting, the guard or shunter would walk back up to No 4 signal box to hand the staff to the signalman. Once shunting was completed, and the locomotive was ready to return to Castle Station, the guard or shunter would trudge back up to No 4 box to obtain the staff to take back to the driver as his authority to proceed

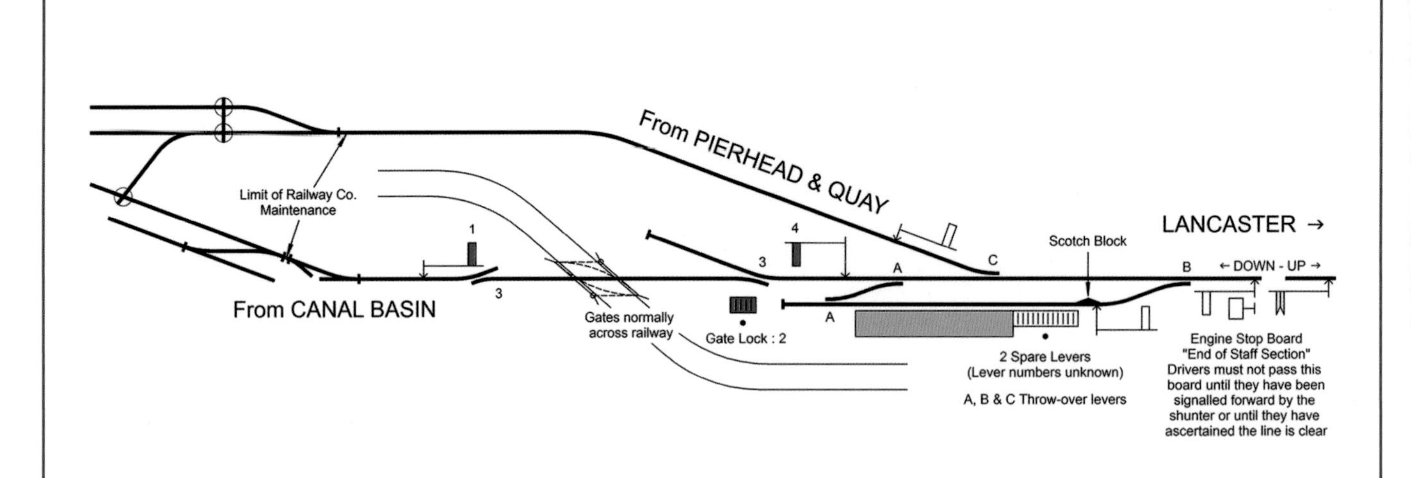
From PIERHEAD & QUAY
LANCASTER →
Limit of Railway Co. Maintenance
From CANAL BASIN
Gates normally across railway
Gate Lock : 2
Scotch Block
← DOWN - UP →
2 Spare Levers (Lever numbers unknown)
A, B & C Throw-over levers
Engine Stop Board "End of Staff Section" Drivers must not pass this board until they have been signalled forward by the shunter or until they have ascertained the line is clear

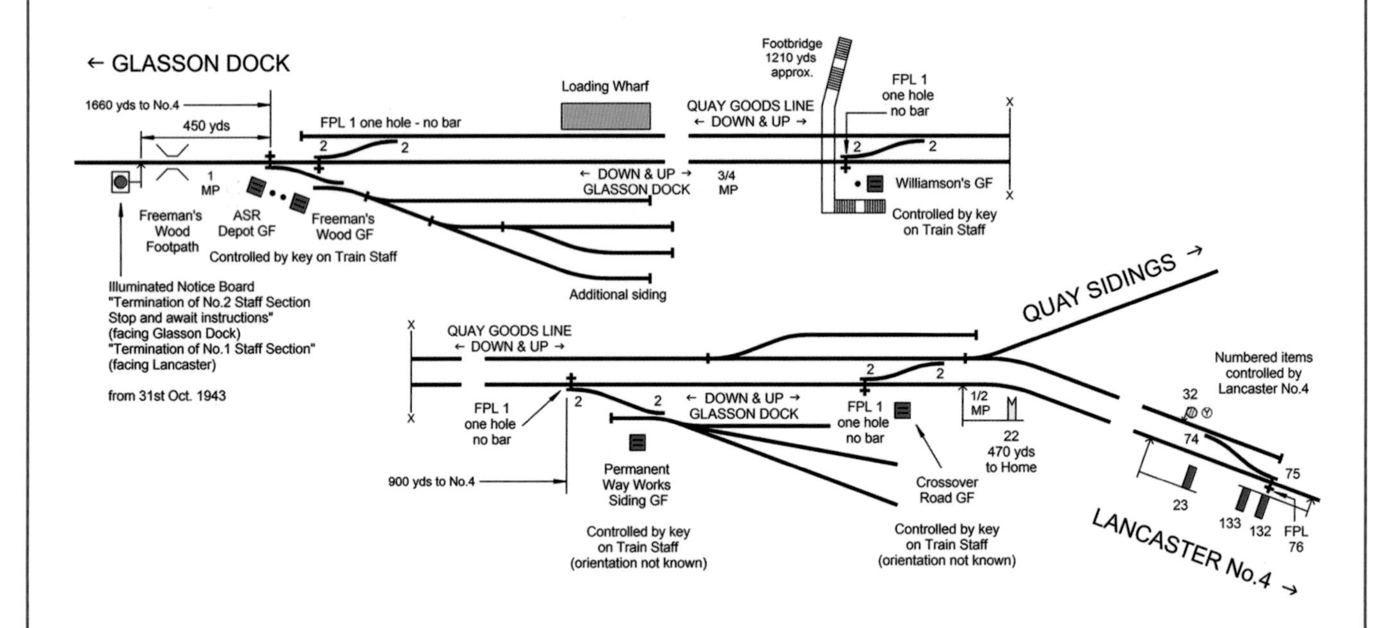
← GLASSON DOCK
1660 yds to No.4
450 yds
FPL 1 one hole - no bar
Loading Wharf
QUAY GOODS LINE ← DOWN & UP →
Footbridge 1210 yds approx.
FPL 1 one hole no bar
1 MP
← DOWN & UP → GLASSON DOCK
3/4 MP
Williamson's GF
Controlled by key on Train Staff
Freeman's Wood Footpath
ASR Depot GF
Freeman's Wood GF
Controlled by key on Train Staff
Additional siding
Illuminated Notice Board "Termination of No.2 Staff Section Stop and await instructions" (facing Glasson Dock) "Termination of No.1 Staff Section" (facing Lancaster)
from 31st Oct. 1943
QUAY SIDINGS →
QUAY GOODS LINE ← DOWN & UP →
FPL 1 one hole no bar
← DOWN & UP → GLASSON DOCK
1/2 MP
22 470 yds to Home
Numbered items controlled by Lancaster No.4
900 yds to No.4
Permanent Way Works Siding GF
Controlled by key on Train Staff (orientation not known)
Crossover Road GF
Controlled by key on Train Staff (orientation not known)
FPL 76
LANCASTER No.4 →

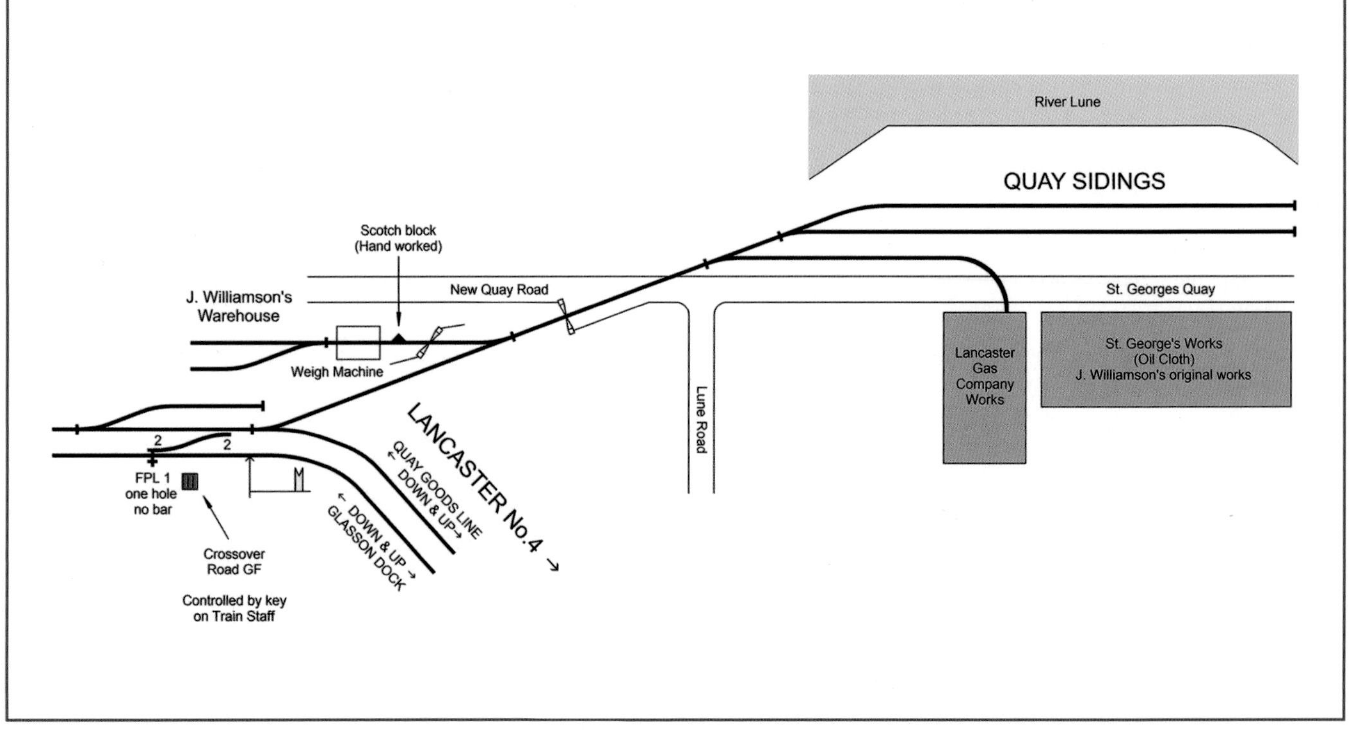
River Lune
QUAY SIDINGS
Scotch block (Hand worked)
New Quay Road
St. Georges Quay
J. Williamson's Warehouse
Weigh Machine
Lancaster Gas Company Works
St. George's Works (Oil Cloth) J. Williamson's original works
Lune Road
LANCASTER No.4 →
QUAY GOODS LINE ← DOWN & UP →
← DOWN & UP → GLASSON DOCK
FPL 1 one hole no bar
Crossover Road GF
Controlled by key on Train Staff

These diagrams show the signalling arrangements as they existed in the later days of the branch's history. The top one depicts Glasson Dock and includes the features that existed between 1932 and closure in 1964. The middle diagrams show the facilities in the vicinity of Lune Mills as they were following the various additions and alterations, which took place in 1943. Finally, the bottom diagram depicts the arrangements on the line to the Ford Quay, following the construction of the siding into Williamson's dispatch warehouse in 1947. Diagrams by Mike Norris.

back up the branch. However, under the rules governing the operation of the train staff and ticket system, the train waiting at Glasson would not be able to proceed back to Lancaster on the ticket alone. The train staff would also have to be present at Glasson station. *In extremis*, it would be necessary to deploy a light engine simply to take the staff to the terminus. Then both locomotives would be able to return to Lancaster. To avoid this scenario, it is likely that trips to the permanent way works siding, wherever possible, were timed to avoid interfering with the scheduled services on the branch.

The arrangements for train control on the Lancaster Quay goods line did not have the complications of the block system. Nevertheless, they were framed so as to provide the maximum flexibility when shunting the traffic to and from Lune Mills and the Ford Quay. Where an engine was already on this part of the branch and, of course, in possession of the train staff, and it was necessary for a second engine to come down to assist with the shunting, the first engine had to return to No 4 box. Here, it would couple up to the second one and the two would return down the goods line together. Where two engines were on the goods line, it was sometimes necessary for one to return to Lancaster, say, with a train of wagons. On these occasions, the returning engine took the staff, together with the fireman of the remaining engine up to No 4 box and handed the staff to the signalman. He, in turn, handed it to the fireman, who now walked back down to the quay lines to hand it to his driver.

The working of the quay goods line had a further complication. We have already seen that, in 1910, a crossover was installed between the Glasson Dock branch proper and the goods line. The two points forming the crossover were locked by the train staffs for the respective lines. Therefore, any driver wishing to use this facility had to have both staffs in his possession. If access to the Ford Quay was also required, the goods guard or shunter would also have needed the key for the padlock on the gate, which closed off the line where it crossed New Quay Road and ran onto the Ford Quay proper. This was quite a bit of hardware for what was, ostensibly, a fairly simple shunting job. In circumstances where a train was already in the block telegraph section on its way to Glasson Dock (and in possession of a ticket only), the arrangements were the same as for access to the permanent way siding, i.e. the signalman at Lancaster No 4 could not hand the driver the staff for the Glasson Dock line until the train out-of-section signal had been received from Glasson Station. The instructions from 1919 also provided for a situation where the engine shunting the quay goods line needed to leave the goods running line blocked with wagons whilst it took a train back up to Castle Station *via* the Glasson Dock line. In this situation, the quay branch staff was handed to the guard or shunter, who remained with the wagons until the engine returned from Castle Station.

Following the withdrawal of passenger services in 1930, the line to Glasson Dock was effectively downgraded and classed as a single goods line. This, in its turn, enabled the LMS to simplify the level of train control on the branch. From 2nd February 1931, the Lancaster to Glasson Dock section was worked on the basis of one engine in steam, thereby allowing the block telegraph instruments to be dispensed with. However, access to the section was still controlled by a train staff. Also, from this date, the crossing keeper roles at the three level crossings on the line were abolished. Instead, the gates were opened or closed by a travelling shunter, who accompanied the train to Glasson Dock and assisted with shunting the various sidings at the station and around the dock. However, for the moment, arrangements on the quay goods line were unchanged with this section remaining a separate operational entity with its own train staff.

In response to a significant increase in traffic on the line during the Second World War, the LMS installed two further crossovers connecting the main running line to Glasson Dock with the long head shunt of the quay goods line. This created what was, in effect, a long loop with an intermediate crossover. In addition, the LMS constructed a new fan of sidings on empty land at the rear of Lune Mills. These wartime additions were controlled by three new ground frames, which were unlocked by a key in the end of the Glasson Dock line train staff. This took the number of separate ground frames at the rear of Lune Mills to five, which, of course, included the frame for the permanent way siding (1905) and for the crossover at the eastern end of the site (1910). As from 31st October 1943, for the purposes of train staff control, the Lancaster–Glasson Dock section was split into two, with a stop board denoting the border between the two sections being positioned a little to the west of the sidings serving the Lune Mills complex. A train travelling to Glasson Dock now had to obtain the staffs for both sections in order to proceed down the branch. However, at the stop board, the staff for the first section from Lancaster would be given up to the shunter and the train would proceed on its way carrying the second staff as its authority to proceed to Glasson. On the return trip, the driver would not be able to proceed beyond the stopboard until he had possession of the staff for the final section up to Lancaster and had been authorised to proceed by the shunter. In this way, a train could proceed to Glasson whilst at the same time allowing a second locomotive to shunt the Ford Quay and the sidings serving Williamson's. Up to this point, the Lancaster Quay goods line had always had its own staff but now this was abolished and the quay line became merely a siding off the main running line to Glasson Dock. This change also removed the necessity for being in possession of both the main branch and the quay branch train staffs when using the crossovers at the rear of Lune Mills. There were two other additions to the layout in the Lune Mills area. The first was the short siding put in at the eastern end of the works and used for discharging tank wagons. The date this was put in is not known but it is likely to have been prior to 1939. The second was the private siding running into Williamson's dispatch warehouse, which was put in during 1947. Both of these had connections to the quay branch and the relevant points were operated by adjacent hand levers.

From here on, there were no further alterations until 7th September 1964, when traffic to Glasson Dock officially ceased. The line was truncated at a point one mile and 143 yards from the junction with the main line. For the moment though, access to what remained of the branch was still controlled by the train staff, and the locking mechanisms on the five ground frames appear to have remained *in situ*. However, on 11th November 1968, the foreshortened branch was reduced to the status of a siding and, as a consequence of this change, the use of the train staff was abolished. This, in its turn, affected the operation of the points and crossovers at the rear of Lune Mills. The ground frames that were unlocked by a key in the end of the train staff were taken out and all the relevant points were then operated from adjacent hand levers. The remainder of what had constituted the Lancaster Quays Branch was closed to traffic on the 30th June 1969 and, shortly afterwards, the connections with what had been the main running line to Glasson Dock were clipped and padlocked pending removal. This left just the stub of the branch with its connections to the District Engineer's yard and what was now Nairn Williamson's power station. Final closure came on 4th April 1971, when the facing connection between the down platform line and the remains of the branch, was secured out of use pending removal.

Chapter Eleven
Locomotives and Rolling Stock

IT IS NOT KNOWN what locomotives were used on the branch in the earliest days. However, photographic evidence indicates that by around 1895, if not earlier, the Webb 0-6-2 Coal tanks held sway. These were a large class of 300 locomotives built in various batches between 1881 and 1897. They were used throughout most parts of the extensive LNWR system and were utilised on goods, mineral and passenger trains. It is highly likely, therefore, that these engines were used on all the trains to and from Glasson Dock and on the quay goods line. Certainly, there is plenty of evidence to suggest that this was indeed the case. There exist at least four photographs taken in the period 1900–1914 showing Coal Tanks in or around Glasson station, including one captured in the act of shunting the Commissioners' sidings alongside the wet dock. Coal Tank No 771, which was turned out in 1885, was delivered new to Lancaster shed and in 1912, engine numbers 580, 948 and 142 are known to have been allocated there. By 1917, the numbers had increased significantly, and the following engines are known to have been present at the shed: 518, 570, 698, 948, 1060 and 2360. However, in LNWR days, if a goods train was a heavy one, for example following a shipment of iron ore, a more powerful tender locomotive might be employed and this is likely to have been one of the 18-Inch Goods Engines or Cauliflowers as they were popularly known. These constituted one of the company's standard classes with 310 examples built between 1880 and 1902. Their use on the trip workings on the quay branch in early LMS days has already been noted. With a heavy train, the locomotive, if necessary, would set back down the branch in order to gain some momentum for the ascent up into Castle Station. These trains would not be allowed to commence this part of their journey until they had a clear route through the station to the yard on the south side of the passenger platforms. In this way, any delays to traffic on the main line would be avoided or at least minimised. An account, from early BR days, of working the heavy linoleum trains up the steep gradient to Castle Station confirms all these features. With a heavy train, the ascent could only be accomplished by building up a good hot fire and taking a long run at the bank. Occasionally, adverse signals or slippery rails in wet weather would thwart even the best preparations. In these instances, the train would set back for another attempt.

The use of LNWR steam railmotors on the branch has already been mentioned. One or more of these units had been transferred to Lancaster shed, probably towards the end of the Great War, and these were put to work on a range of local duties. They seem to have first appeared on the branch at some point between 1921 and 1923. Certainly, in the aftermath of the Great War, there does seem to have been an attempt to significantly improve the frequency of the service on the branch and the use of the railmotors was perhaps seen as an economical way of doing this. By 1923, they were handling most, although by no means all, of the passenger services to and from Glasson. The LNWR had constructed six of these units in 1906–07 with one further example built in 1910, making a total of seven. They combined the accommodation of a passenger carriage with a steam engine, the latter being situated in a separate compartment at one end. Both were placed on a single chassis within a carriage type, panelled body. They could be driven from either end, so there was no need to use the run-round loop at Glasson Station. The driver simply walked to the other end, rather in the manner of a modern diesel or electric unit, while the fireman, on the other hand, always stayed at his post in the engine compartment. The passenger accommodation was restricted to third class only, arranged in an open plan format, described perhaps euphemistically as "one class only". On the branch, it was recalled that almost by tradition, male passengers sat at one end of the single compartment and women at the other, with inevitably perhaps, the children running up and down the centre gangway. Unfortunately, no photographs have so far come to light showing one of these units running between Lancaster

LNWR Coal Tank No 771 shunting at Lancaster in the early years of the twentieth century. This particular locomotive is known to have been allocated to Lancaster shed in 1885. She is very likely, therefore, to have been a regular performer on the goods and passenger trains to Glasson Dock and on the trip workings on the Lancaster quay goods line.
CRA, LCR Collection.

Another version of *Engineer Lancaster*. This one is a member of the Samson class and had originally been No 414 *Prospero*, which was built in 1873. She was the District Engineer's locomotive from 1903 until withdrawal in 1924. The photograph was taken at Lancaster engine shed with the grounds of the Royal Albert Asylum in the background.

Courtesy Lancaster City Museums.

and Glasson. Locally, they appear to have been nicknamed the *Puff and Dart* by the travelling public. The railmotors were intended to provide limited passenger accommodation at a lower cost than a conventional locomotive with a rake of carriages and, for this reason, they tended to be deployed on the more lightly used branch lines. Indeed, on the line between Garstang and Knott End, just a little to the south of Lancaster, these railmotors virtually monopolised the passenger service from 1920 until it ceased in 1930. However, this type of carriage and locomotive combination tended to suffer from a lack of power and would begin to struggle if goods wagons were attached to the rear or high numbers of passengers suddenly presented themselves and it became necessary to attach an additional carriage. This is possibly the reason why, in 1923, these units were conspicuous by their absence on the busiest early morning trains and on the late Wednesday and Saturday-only services. Although the limited single class accommodation may also have been a factor. However, Mr Gilmour recalls that, by 1927, all the passenger services on the line, with the exception of the extra trains on Saturday were worked by the railmotor. On these services, a locomotive hauling two elderly six-wheel carriages deputised for the railmotor. Mr Gilmour recalls that one of these still retained its LNWR plum and spilt milk livery. There was seating in the railmotor for 48 third class passengers, however, an LNWR document from 1919 indicates that the two-carriage set in use on the branch at that time had a seating capacity of 66 — just the job for the last train from Lancaster on a Saturday night. There is a story that one of the railmotors broke an axle coming onto Lancaster shed and was derailed. A breakdown crane was summoned from Preston. However, the weight of these units was very unevenly distributed, given that the engine compartment together with the coal and water supplies were situated at one end. No account of this was taken when the crane carried out the lift and consequently the chassis simply bent in the middle, the damage being such that the railmotor was deemed to be irreparable. These slightly unusual self-propelled passenger vehicles disappeared from the branch around 1928–29 and for the remaining period, the entire service was operated by a conventional locomotive with a train of three six-wheel carriages.

The use of ex-LYR 0-6-2 tank engines on the branch in the years immediately after the grouping has already been discussed. However, by the later 1920s, ex-LNWR 5ft 6in 2-4-2 tank engines were beginning to put in an appearance on both goods and passenger trains. There is a fairly well-known photograph of No 6620 at Lancaster coming off the branch with a long goods train composed almost entirely of private owner coal wagons. There is also a distant view of what appears to be a member of this class hauling a train of three, six-wheeled passenger carriages near Conder Green. Glasson Dock's last stationmaster, Mr EH Gilmour, recalls that following the disappearance of the railmotors, most of the passenger services on the branch were worked by these ex-LNWR 2-4-2s from Lancaster shed. However, there were a couple of exceptions. He states the first passenger working of the day, the 7-10 am from Lancaster and the corresponding return train from Glasson were worked by the District Engineer's departmental engine, *Engineer Lancaster*. In the late 1920s, this was a member of the ex-LNWR Waterloo Class, more commonly known as Small Jumbos and had originally been No 737 *Roberts*. As we have already seen, these locomotives were kept at Lancaster shed, but remained an asset of the District Engineer's department. It is, therefore, difficult to say whether this was a regular or an infrequent occurrence. The other exception to the diet of 2-4-2s was the 5-00 pm from Lancaster and its return working from the terminus. Mr Gilmour recalls that this was in the hands of an ex-LNWR 4-4-2 Precursor tank from Oxenholme shed. This is entirely plausible, but again, it is not possible to say to what extent this was a regular occurrence.

The small shed at Lancaster was situated to the south of Castle Station in the vee of the junction between the main line and the Old Station Yard. It was a sub-shed of Preston and, in 1925, had eleven locomotives under its care. For many years the shed supplied the motive power for the services on the branch until it was closed on 4th February 1934. The staff and some of the locomotives were transferred to Green Ayre shed, which was the old Midland Railway establishment adjacent to the station of that name. At the time of closure, the only goods engines allocated to the Lancaster LNWR shed were two Fowler 0-6-0 tanks, numbers 16464 and 16609. These must have been frequent performers on both the daily trip workings on the quay goods line and on the irregular forays out to Glasson Dock. There are a small number of photographs taken at the sidings at the rear of Lune Mills around 1945 and these show this class of engine at work. In the post-war period, Fowler 4F No 4032 was a regular on the branch, being photographed on the Ford Quay and at Glasson Dock, first of all carrying its LMS number and then later on in its BR guise as 44032. The locomotive is known to have been allocated to Green Ayre shed until November 1954. She is recalled charging up the bank from Williamson's with a train of vans carrying linoleum and then running south through platform three at Castle station. In his book *London Midland Fireman*, Mike Higson heaps praise on 4032 saying that she was half as strong again as any other 4F and was the only one of the class at Green Ayre that could bring a full load of wagons up the bank successfully. On occasions there might be a train composed of as many as 40 loaded vans to be brought up the branch from the siding serving Williamson's warehouse. Other members of this class seen on the branch were Nos 43890 and 43984. By the 1960s, the Ivatt-designed 2-6-0 2MT locomotives seem to have been the most common performers. In particular, Nos 46422 and 46433 appear in a number of photographs taken on the branch at

Fowler 4F No 4032 still sporting its LMS livery and number is captured in the course of shunting the sidings on the quayside at Lancaster. This engine was a regular performer on the branch and was allocated to Green Ayre shed until November 1954. The photograph was taken on the 25th May 1951. CRA, Pearsall Collection.

A stunning portrait of Ivatt 2MT No 46422 standing on the grass-grown line to the wet dock. In the background on the left are the twin storage sheds erected by the LNWR shortly after the opening of the line. The sailing vessel at one of the river berths is the *Moby Dick,* which spent a little time at Glasson Dock before moving to Morecambe to become a permanent floating exhibit. Taken on the 2nd March 1962. Ron Herbert.

Ex-LNWR 5ft 6in 2-4-2 tank No 6620 brings a train composed mainly of coal wagons off the Glasson Dock branch and into Castle Station. This class of engine seems to have been allocated to the services on the branch from around 1925 and there are photographs showing them engaged on both goods and passenger work.

L&NWRS.

this time. However, there are also images showing at different times, ex-LMS Fowler 0-6-0 tank No 47662, and a Fairburn-designed 4P 2-6-4 tank engine, No 42063. The former was engaged in shunting the quay branch whilst the Fairburn tank was working one of the nitrochalk specials.

Turning now to the types of carriage used on the passenger trains. Shortly after the line opened to passengers on 9th July 1883, the LNWR's Traffic Committee ordered a single coupled train for the branch. The term "coupled" was used by the railway to describe a train arranged in a specific formation and used on a particular route. An undated LNWR document, probably from the first years of the twentieth century, provides details of coupled trains on the Lancaster and Carlisle Division. This reveals that the Glasson Dock branch was served by two, three-coach sets of six-wheel carriages. Both sets were configured thus: brake third, composite (first and second class), brake third. However, within this, there was some variation in how the compartments in the composite carriages were arranged and in the amount of space taken up by the guard's compartment in the brake thirds. All these carriages were gas lit and unheated. Although the dates of construction of the individual carriages are not recorded, it is likely that these were elderly vehicles coming to the end of their lives. This was usually the practice on minor routes like that to Glasson Dock. As older stock was displaced by more modern vehicles, it tended to be cascaded down, firstly to secondary services and then finally to branch lines. By 1906, the branch had been assigned a single, three-coach coupled train, composed of 30ft 6in six-wheeled carriages, once again in the brake third, composite, brake third configuration. This particular length of carriage dated from the 1870s and so again, these vehicles would have been fairly elderly at this point.

An LNWR document from June 1909 concerning the conversion of second class compartments to third on certain branch line trains reveals more about the carriages used on the Glasson Dock branch. At this point, there were once again two coupled sets, each composed of three carriages. Set one was configured thus: brake third, luggage composite (first and third), brake second. In both of the brakes, the guard's compartment was situated at the outside end of the carriage. Set two was composed of a brake third, a luggage composite (all three classes) and then a composite brake again with accommodation for all three classes. In this set, the guard's compartments in the brakes were flanked by passenger compartments.

The Lancaster Port Commissioners' minutes show that for several months from the late Summer of 1909, the LNWR abolished first and second class on the branch passenger services. The minutes suggest that this was achieved by the simple expedient of removing the composite carriage from the branch set. However, this would not have been possible with the sets configured as they were in June 1909 above. This alteration seems to have elicited some complaints from the better-off and, therefore, more influential members of the travelling public, and there was an exchange of letters between the Commissioners and Mr Price on the subject. In the end, the railway company capitulated, and first and second class accommodation was reinstated, probably in October of 1909.

By 1912, an LNWR diagram of coupled trains reveals that once again there was now only one set covering the entire passenger service in the branch. This was composed of two six-wheeled carriages, one being a 32 ft long brake composite (all three classes) and the other a 30ft 1in brake third. In both cases the guard's compartment was flanked by passenger compartments. As before, this coupled set was gas lit but still not fitted with any form of heating.

An LNWR diagram of local carriage working dated 1st October 1919, reveals a broadly similar arrangement. However, second class on the LNWR had been abolished with effect from 1st January 1913, so the carriages now have only either third- or first-class compartments. The branch's single coupled train still consists of only two carriages. The configuration of the set and the arrangement of the individual compartments within the carriages is not revealed. However, the train as a whole, provided seating for 60 third class passengers and six first class, which suggests a single first-class compartment in one of the two carriages.

The Cumbrian Railways Association holds part of what appears to be a slightly later version of this document and includes details of the carriage arrangements for the Glasson Dock branch. There are now two sets, the first one covering the morning services and the second the afternoon trains together with the additional services on Wednesdays and Saturdays. The latter were brought in shortly after 1919. The document indicates that the entire passenger service was covered by conventional carriages, so this seems to be prior to the introduction of the railmotor onto the branch. This narrows down the time frame to between 1920 and 1922 inclusive. In addition to their runs up and down the Glasson Dock branch, both sets of carriages spent some of their time on Lancaster–Morecambe services, which, conveniently, left the former place from the same bay platforms as the Glasson Dock trains.

As we have already seen, in the later 1920s, a two-carriage set of six-wheelers deputised for the railmotor on the late Saturday evening trains. Once the latter was taken off the branch passenger service around 1928–29, the trains were worked by a three-carriage set of six-wheelers. The configuration of this is not known. There exists a rather distant view of what is probably this set, seemingly hauled by an ex-LNWR 5ft 6in 2-4-2 tank engine. However, the image is too blurred to attempt any identification of the individual carriages.

Acknowledgements

Although my name appears on the front cover of this book, I am very mindful that in reality, any publication of this nature is the result of a happy collaboration between a number of individuals. Without their willing help it would have been very difficult, not to say impossible to pull everything together and bring the project to a successful conclusion. Whilst I am delighted to acknowledge the assistance provided, I am also happy to state that any errors appearing within these pages are wholly mine.

I must especially thank Mike Norris of Lostock Hall who very kindly shared with me his research on the signalling on the branch. He also provided the excellent signalling diagrams accompanying this book. Mike also acted as a sounding board for some of my ideas and theories especially those relating to some of the slightly mysterious things that happened on the branch during the Second World War. I will cheerfully acknowledge that signalling is not one of my strongpoints, although, hopefully I am getting better and Mike's assistance has been invaluable.

Special thanks also go to author and retired railwayman Ken Nuttall of Lancaster who, over a number of visits to his home, freely made available to me his extensive collection of photographs and documents and shared with me his knowledge of the branch and of railway matters generally.

I would also like to thank Ron Herbert and Derrick Codling for allowing me to use their photographs in this book. I have bracketed these two gentlemen together as on at least one occasion in the 1960s they accompanied one another on a photographic excursion onto the branch. Consequently, some of their images cover the same event but from a different perspective. Taken together, the two groups of photographs provide an evocative portrayal of the branch in the early 1960s. Ron and Derrick also provided a number of insights into the operation of the branch during this period.

Thank you also to Guy Wilson, the archivist of the Cumbrian Railways Association, who scoured his cupboards and drawers to provide me with a copious supply of documents relating to the branch and to railway operations in the wider Lancaster area.

During the course of gathering material for this book I made frequent visits to Lancashire Archives in Bow Lane, Preston. I spent many hours going through the records of the Lancaster Port Commissioners and those of James Williamson & Son, the floorcovering manufacturers. I would like to place on record my sincere thanks to the search room staff there for their unfailing professionalism and helpfulness. In a similar vein I would also like to say thank you to the people at Lancaster Museums, in particular Naomi Parsons and Richard Whittaker for allowing me to use items from their wonderful collection of photographs. This provides extensive coverage of the history of the Lancaster area and also of the city's rich maritime heritage. Many thanks also to the very helpful and friendly staff at Kendal Archive Centre. I would also like to thank the following individuals: Richard Foster, Ian Lydiatt, Philip Millard, David Patrick, Howard Quayle, Stewart Shuttleworth, Pete Skellon, Russell Wear and Bob Williams.

Finally, I would like to pay tribute to the members of the Cumbrian Railways Association's Publications Team for their hard work in putting this book together. First of all, to Mike Peascod, the Publications Manager, for converting my words and the photographs into a publication fully up to the Association's high standard. Also, to Alan Johnstone and Philip Grosse who, between them, have produced all the wonderful maps and architectural drawings that grace the pages of this book. Philip was also at one time an employee at Lune Mills and provided valuable insights into the activities of James Williamson & Son Ltd and Nairn Williamson Ltd in the 1960s.

Ladies and gentlemen, thank you.

Principal Sources Consulted

Bibliography

A History of Lancaster, Editor Andrew White, Edinburgh University Press 2001. ISBN 0 7486 1466 4.
An Illustrated History of LNWR Engines, Edward Talbot, OPC 1985. ISBN 0 86093 209 5.
Beeching, 50 Years of the Axeman, Robin Jones, Heritage Railway 2011. ISBN 978 1 906167 68 4.
Glasson Dock: The Survival of a Village, John Hayhurst OBE, Centre for North West Regional Studies Lancaster University, 1995. ISBN 0901 800848.
Glimpses of Glasson Dock & Vicinity, Ruth Z Roskell, Landy Publishing, 2005. ISBN 1 872 895 34 4
Lancaster's Little Ships, Edward Gray, privately published, 1996. ISBN 0 9529643 0 9.
Lord Linoleum, Philip J Gooderson, Keele University Press, 1996. ISBN 1 85331 146 4.
LNWR 30ft 1in Six Wheeled Carriages, Phillip Millard, LNWRS, 2008. ISBN978 0 9546951 6 3.
North of Leeds, Peter Baughan, David and Charles 1966.
North of Preston, Peter Baughan, Unpublished manuscript in the possession of the Cumbrian Railways Association.
Over Shap to Carlisle, Harold Bowtell, Ian Allen Ltd 1983. ISBN 0 7110 1313 6.
Private Owner Wagons: A Thirteenth Collection, Keith Turton, Lightmoor Press 2014. ISBN 13: 9781899889938.
Private Owner Wagons: Volume One, Bill Hudson, OPC 1976. SBN 902888 70 6.
Railways Around Lancaster, K Nuttall and T Rawlings, The Dalesman Publishing Co Ltd 1980. ISBN 0 85206 578 7.
Storeys of Lancaster, Guy Christie, Collins 1964.
The Canals of North West England, Vols 1 and 2, Hadfield and Biddle, David and Charles 1970. ISBN 0 7153 4956 2.
The Lancashire and Yorkshire Railway: Vol 1, John Marshall, David and Charles 1969. ISBN 7153 4352 1.
The Lancaster Canal in Focus, Janet Rigby, Landy Publishing 2007. ISBN 978 1 872895 72 7.
The Last Tide: A History of the Port of Preston: 1806–1981, Jack Dakres, Carnegie Press 1986. ISBN 0 948789 02 6.
The Leaves We Write On, Mark Cropper, Ellergreen Press 2004. ISBN 0 9549 191 1 4.

Newspapers and periodicals:

Lancaster Guardian, Lancaster Gazette, London Gazette, Morecambe Guardian.
British Railway Journal Spring 1987, Stanley C Jenkins MA. The Glasson Dock Branch.
British Railway Journal Summer1987, 'Operating the Glasson Dock Branch'. Mike Christensen.

Lancashire Archives:

DDLPC Lancaster Port Commission 1744-1981. Various.
DDX 909 James Williamson & Son, Lancaster. Various.
DP432/56 Deed and tenancy plan for the Lancaster and Carlisle line. Undated.
DP423/60. Plan of St George's Quay. Undated.
DP432/61. Map of complete branch. Undated.

Kendal Archives Centre:

WDSo 108/A 3135/Bowtell 25. Notes on the Glasson Dock Branch.
WDSo 108/A 6. Notebook containing chronological list of notable events on the Furness Railway. Contains notices relating to Royal train working from Glasson Dock 1917.

The National Archive:

AN155/90 BR file relating to closure 1963-1971.
RAIL 410/64/86 Chairman's memo 1878.
MT 6/343/1 BOT inspection 1883.
MT 6/1399/8 BOT inspection 1905.
MT 6/1936/5 BOT inspection 1910.

Rear Cover:
These three photographs provide further commentary on No 42301's final journey to Glasson Dock for scrapping. By this time, any traffic for the terminus was usually handled by one of the daily shunting turns serving the Lancaster Quay goods line. Consequently, the dead locomotive was brought down onto the quayside at Lancaster and shunted with the rest of the wagons in the train, before making the final journey to Glasson. Indeed, the image on the front cover captures both locomotives visiting the District Engineer's yard. The first photograph on the rear shows 42301 and 46422 on the quayside with the River Lune in the background. Although not visible in these photographs, No 42301 was accompanied to Glasson Dock with a rake of fitted vans, which appear to have come from the dispatch warehouse at Williamson's Lune Mills. Upon arriving at the terminus some shunting was carried out to place the brake van between the withdrawn locomotive and the vans. In the middle image, No 46422 is running around the train for the final time. The bunker of 42301 is just visible on the right-hand side. In the final image, the locomotive is being propelled through the gates separating BR metals from the siding serving the ship-breaking facility of Lacmots Ltd. With the dead locomotive placed in the sidings, No 46422 hauled the remainder of the train back to Lancaster. All three photos, Ron Herbert.